The Changeless Order

THE PHYSICS OF SPACE, TIME AND MOTION

The Changeless Order

THE PHYSICS OF SPACE, TIME AND MOTION

Edited with Introductions by

ARNOLD KOSLOW

Assistant Professor of Philosophy
Brooklyn College (C.U.N.Y.)

GEORGE BRAZILLER
NEW YORK

TO MY PARENTS

Charles and Yetta

For information, address the publisher:
George Braziller, Inc.
One Park Avenue, New York, N.Y. 10016

Library of Congress Catalog Card Number: 67–24207
First Printing
Printed in the United States of America

Preface

THIS VOLUME IS INTENDED to supply a basis for understanding and appreciating a coherent set of issues which are central to physics. These issues are represented by the writings of only those authors who have either pioneered or greatly shaped the results which they describe. Physics, however, is one science among many, and even within it there are sciences such as Optics, Thermodynamics, Atomic and Nuclear Physics, as well as exciting boundary sciences like biophysics and quantum chemistry, which have not been represented. The task of elucidating the problems and character of all those sciences within and neighboring upon physics is beyond the scope of a single volume.

The problems represented center around the description and explanation of the motion of middle-sized, non-atomic bodies. We have therefore included much of the scientific deliberation over the adequacy of various concepts of space, time, and matter, since these concepts have figured largely both in descriptions and explanations of motion. Although Einstein's Special and General Theories of Relativity have been represented, there is no discussion of the philosopher Alfred N. Whitehead's Theory of Relativity, nor that of Professor E. Milne. Neither have we included Professor V. Fock's critique of the Einstein General Theory. These critiques and alternate theories have not entered sufficiently into the mainstream of scientific discussion to be included, although no serious student of the subject can afford to neglect them.

The Law of Inertia is the subject of a good deal of discussion in this volume, chiefly because it has a fundamental role in the explanation of motion. There are other laws of physics no less

v

important, such as Newton's Second Law of Motion, or his Law of Universal Gravitation. But the scientific discussions which have centered around the latter, for example, though important because they raised the issue whether action-at-a-distance is an intelligible type of explanation in physics, do not seem as central or as deep as those difficulties centering around the Law of Inertia.

The Law of Inertia is a law of conservation, and we have stressed the importance of this kind of law for physics. But there are other types of laws which state how certain physical quantities and structures *change* in the course of time. Newton's Second Law of Motion, and Schrödinger's Equation in Quantum Mechanics, provide, under suitable circumstances, information about the way in which the state of a system changes in time. These laws of development are no less characteristic of physics than laws of conservation, nor are they less important, for they usually constitute the backbone, the strongest premise, of a scientific explanation. However, the laws of conservation are more closely linked to the problems of space and time so that their inclusion renders the total set of represented issues more cohesive.

We have excluded any extended discussion of Quantum Mechanics, for two reasons. Quantum Mechanics raises a serious challenge to the view that all theories of physics are causal or deterministic. The debate between A. Einstein and N. Bohr, whether a satisfactory theory of atomic phenomena should be causal, is one of the great debates of modern physics. However, to place the problem of causality in perspective requires a book, rather than one or two selections. Quantum Mechanics has also presented a challenge to the usual ways of speaking about and representing the motion of bodies in a spatio-temporal continuum. The discussion of this issue is exciting and vital, but it still seems to be beyond the reach of a clear non-technical statement.

I would like to thank G. Feinberg, S. Morgenbesser, and E. Nagel for their timely and helpful advice. My special thanks to my wife Susan, who helped construct this volume, literally and figuratively.

March, 1967   *Arnold Koslow*

Contents

Introduction

No ONE BOOK CAN capture everything of significance that the physical sciences have had to say. Nevertheless there are certain concerns which are central and typical, and we have tried here to bring together original sources and contributions which bear on a few of those issues. The selections in the book are organized into two sections: Part I, Motion, and Part II, Conservation.

The motion of bodies seems to be one of these central scientific concerns. Astronomy, the earliest, and classical mechanics, perhaps the most exact of the physical sciences, were concerned with the description and explanation of the motions of bodies. A rich variety of concepts of space, time, and matter were employed in framing some of the laws of motion, one of the most important of which is the Law of Inertia. Some of the readings in Part I are devoted to the various concepts of space and time relevant to specifying the Law of Inertia in a cogent way; other readings indicate changes in it. Indeed, changes in the Law of Inertia suggest a way, which we will describe below, in which the physical sciences have been cumulative. Part I concludes with a piece on the explanation of the Law of Inertia.

Part II contains some attempts to find certain constancies in nature, a search which is probably as old as disputations about the Real. Certainly, conservation was part of Plato's argument that what is most real is that which is constant, unchanging, and eternal.

The kinds of constancies sought and found by scientists vary widely in type and importance. Some principles of conservation were believed to be more fundamental than others, and all re-

1

quired organization, explanation, and support. The principle that mechanical energy is conserved under special conditions was explained by Helmholtz, but proved less fundamental than a principle which said that mechanical energy plus heat was conserved under still more general conditions. Today it is believed that energy in all its forms is conserved, although this principle has no current explanation.

It has also recently been discovered that the principles of conservation of the various forms of energy are closely related to certain principles of spatial and temporal symmetry. We have, therefore, included articles which explore the validity of the major principles of conservation and explain how they are intimately linked to our concepts of space and time. The two parts of this book therefore form a coherent whole. They by no means represent all the major issues of physical science, but we hope that even this connected set of problems will convey to the reader a sense of the multifaceted concerns and achievements of the physical sciences.

Two Views of Science

While it is the aim of a collection such as this one to point up the aims and record some of the achievements of the physical sciences, there is an issue to which the reader should be alerted. For he may believe that science is identified either with some special method, the so-called scientific method, or with some special body of truths—the scientific truths.

Often it is argued that the birth of modern science took place in the seventeenth century, because it was then that scientists first required that their statements be sufficiently clear and definite so that they might be given unequivocal tests. Seventeenth century scientists, according to this view, agreed to reject their theories if they proved false. The fact that we know that some seventeenth century theories are incorrect, that we know more than those scientists knew, is, according to this view, relatively unimportant for our understanding of what is scientific. It is to the scientific method that we ought to look for an understanding of the scientific undertaking, and this method has been substantially the same for roughly three centuries.

On the other hand, a more detailed look at some of those

earlier scientific works can easily lead to an alternate understanding of science. Side by side and within these scientific works, we find firmly held beliefs about witchcraft and the possibility of supernatural phenomena, references to the action of God to explain the constant amount of energy in the universe, and disquisitions on the yearnings which the planets have to approach the sun. Such considerations suggest that surely the content of certain beliefs should enter into a consideration of whether something is scientific. Such a view suggests that a science is an organized body of knowledge which is empirically meaningful. Great care has gone into the specification of this view. The content of the statements which constitute this body of knowledge is supposed to be regulated or governed by a criterion of empirical meaningfulness, and the organization of this knowledge is along deductive or inductive lines. Further, it has been argued that these bodies of knowledge explain, predict, or in other ways support and systematize true statements about the world. We shall not go further into the problems of clarifying these terms nor shall we review the various theses about how the sciences explain, predict, or systematize, but it should be noted that this view of science does not require science always to consist of the same body of statements organized in the same way at all times. The structure of a science may undergo reorganization and, over time, certain statements may be accepted and others rejected from the scientific corpus.

There are, then, two prominent polarized views about the nature of science: one identifies science with a method and claims that there has been only one scientific method since the seventeenth century. The other view claims that science (or a science) at a given time is a body of knowledge organized along special lines. We have stated these positions in a rather bare, unqualified form. Nevertheless, from either view, there seems to follow one important consequence: that no particular significant sentence can be judged scientific when considered by itself. For example, according to the second view, we would have to know whether that sentence was part of a scientific corpus satisfying certain conditions. It would be a mistake to read the selections in this book, trying to determine whether each statement is scientific. Thus, if the reader scans these writings with the idea that there is one single method exemplified by all these writers, then he will be engaged in a task for which these works can provide little help. On the other hand, if the reader looks for a body of truths about the world, organized into a certain structure of laws and theories, then I think he will find himself frustrated by the wealth of divergent material.

The Cumulative Character of Science

In this book we have also had the subsidiary but related aim of showing how the sciences have been cumulative, despite the fact that many statements once firmly held have been shown false, and despite the fact that there is no apparent aim on the part of these scientists to institute, describe, or enforce any one scientific method which all scientists follow or ought to follow.

The physical sciences are cumulative in many ways. If a science is at least a body of knowledge, that is, statements known to be true, then an increase of the body of knowledge during the course of time can make a science cumulative in a numerical way.

Many people believe that the scientific body of knowledge should include not only those statements known to be true, but also those statements for which we have good grounds for believing true. The difference is mainly that if a statement is known to be true, then it is true. But if we have a good basis for believing a statement, that statement may nevertheless be false. Let us call a statement presumably true if it is either known to be true or if we have good grounds for believing it true. If we widen the scope of scientific knowledge so that a body of knowledge includes all statements which are presumably true, then science fails to be cumulative numerically. For it is false that once a statement is accepted into the corpus of science, it will never be rejected. The proposition that all terrestrial bodies have a tendency to move toward the center of the earth, where they will forever remain at rest, is no longer (pace Aristotle) presumably true.

The sciences are also cumulative once it is required that the acceptance of a statement into the scientific corpus depends upon the relation of that statement to the total body of knowledge at that time. Without such a requirement, seemingly correct rules of acceptance would have to be abandoned. Suppose, for example, that we agreed to accept a statement S if S were confirmed to a high degree by some statements already accepted. Suppose that we know that 99 percent of all Texans are millionaires (to borrow an example of Carl Hempel) and 99 percent of all Texan philosophers are paupers. The first statement together with the information that Smith is a Texan is apparently good grounds for believing that Smith is a millionaire. On the other hand, suppose that Smith is also a philosopher. This fact, together with our information about the financial status of Texan philosophers, is good ground for believing that Smith is a pauper. However, if we accept into a body of knowledge whatever statements we have good grounds for believing, then we have accepted two incompatible statements about Smith: that he is a millionaire and he is a pauper. This paradox suggests that whenever we determine to accept a statement on the basis of the evidence, we should do so on the basis of *all* the evidence for it. The total evidence requirement just described seems to be a reasonable one which all rules of acceptance ought to satisfy.

There are, of course, other ways in which physical sciences are, or ought to be, cumulative. It is an obvious fact that certain theories are taught to physics students before others: classical mechanics, for example, is generally taught before relativistic mechanics. One reason for such priority has been that it is easier to learn the relativistic theory after mastering the classical one. We do not know whether there is sound psychological evidence for this claim, but the reason offered does suggest that there is perhaps some kind of psychological cumulativeness, a heuristic ordering among the theories of the physical sciences. Related to this suggestion, but very different in character, is the observation that the physical sciences have explained various phenomena and in turn have explained or sought to explain the premises of their explanations. It must be noted that even if the premises of an initial explanation have been explained, the initial demonstration does not cease to be explanatory. For example, Newton's Law of Universal Gravitation can be used to explain why the earth moves on an elliptical orbit. An explanation of the Law of Gravitation would not cancel the first explanation. The first explanation would be less fundamental, if we had the second, and to that extent it may have less worth. But it remains an explanation nevertheless. Scientists have been able, therefore, not only to add to their stock of explanations, but also to explain their explanations, thus enriching science in a vertical as well as a horizontal dimension.

Related to, but not identical with, the growth in explanatory power is the increased accuracy of the physical sciences. And this might suggest that accuracy be taken as a measure of the growth of the sciences. However, it would be a mistake to take an increase in accuracy as an invariable sign of scientific growth. No one would put much stock in increased accuracy beyond the point where it makes no difference to current or imminent investigations. Also, many would agree that Quantum Mechanics represents an advance over classical Newtonian Mechanics, although in certain respects there has been a loss in accuracy. For example, given the exact position of a particle at two different times, the classical laws of mechanics were such that the exact velocity of the particle could be predicted for any future time; but the laws of the quantum theory do not permit any such prediction.

None of the five types of cumulativeness we have mentioned thus far can be unequivocally attributed to the course of the physical sciences without further qualification. We could not seriously maintain, for example, that the history of the physical sciences shows that scientific statements have become more exact. To these five types we wish to add yet a sixth. This type of cumulativeness may be conveyed by taking as an example a law such as the Law of Inertia. In the form which Newton gave to it, it states that if no forces act on a body, then that body remains at rest, if at rest, and if in motion, then it continues to move in a straight line with constant velocity. Kepler believed in a different Law of Inertia: if no forces act upon a body, then if the body is at rest it will remain at rest. The two statements are not equivalent, yet there is the important fact that both statements were believed at different times to be the Law of Inertia. Why each statement was thought to be the Law of Inertia is a delicate matter having to do with the role that the Law of Inertia is supposed to have in an explanatory scheme. We shall say more about this in the Introduction to Part I. But Kepler and Newton offered generalizations which were supposed to fulfill the explanatory role required of a law of inertia. In this way, the various attempts to fill in the explanatory scheme with theoretically and empirically adequate statements may be reckoned as cumulative.

Inertia is not a singular case. The Kinetic Theory of Gases and the Law of Conservation of Energy are laws or theories which have been modified and corrected during several centuries of scientific activity. They, too, have an invariant sense or meaning despite the changes which have taken place in them, so that we understand these changes as contributions *to*, or changes *in*, the Kinetic Theory of Gases, changes *in* the Law of Inertia, changes *in* the Law of Conservation of Energy.

The physical sciences have been cumulative in all these dimensions. But the last sense has been wrongly neglected in recent discussions of the sciences. We find in the history of the physical sciences that there have been series of contributions to, and changes within, the same scientific theory. And this strongly suggests that such theories have an invariance of meaning despite the changes of modifications in them.

We have selected material which is of scientific and philosophical import, underlining more than is usual the cumulative aspect as well as the positive achievements of the physical sciences.

PART I ooo

Space, Time and Motion

Aɴʏ ᴛʜᴇᴏʀʏ ᴏꜰ ᴍᴏᴛɪᴏɴ requires the specification of some concept of space and time. The first part of this volume, therefore, contains a number of readings designed to give a profile of the variety of concepts of space, time, and matter which have been employed in scientific description and explanation. We have tried also to catch something of the interplay among the partisans of these descriptive and explanatory schemes. Newton's theory of Absolute Space and Time as well as the so-called relational theories of Descartes and Leibniz are included. Certain key works such as Euler's defense of the Newtonian viewpoint and Kant's influential theory of space and time, which was neither absolute nor relational, are also represented. Theories like those which distinguished sharply between matter and space and those like Descartes', which identified the two, are contrasted with the theory of Mach, who denied the existence of matter. The conflict between the theories of Plato and Newton, who attributed causal efficacy to space, and those who did not are documented. Newton's discussion of the experimental proof of rotation with respect to Absolute Space and Mach's famous rebuttal of it have been included. Einstein's descriptions of his revolutionary proposals which successfully challenged and radically changed all precedent concepts of space, time, and matter are also given.

We have also made our selections wide enough to include a discussion of one of the most fundamental laws of physics: the Law of Inertia. This law, as will be seen, cannot be discussed independently of the concepts of space and time. There has been such a rich variety of laws of inertia that the reader will find in

9

these selections examples of scientific discussion and scientific change at their best. In the remainder of this introduction a coherent scheme of the variety of views which the reader will encounter is described.

Although every author seems agreed on the necessity for some concept of space, place, and time for the business of physics, no universal accord is clearly evident as to which of these concepts should be used. One idea, however, which seems to have been shared by most scientists, is that under *normal* conditions bodies will execute *natural* motions. This statement about bodies, if true, is called the Law of Inertia. Throughout the history of science there have been differences over what constitutes a normal condition and what kind of motion is natural. Despite these differences there was the shared assumption that only one kind of condition was normal and only one type of motion was natural. It follows from the Law of Inertia stated above that if a body moves in any way other than the natural one, then the conditions under which it is moving are not the normal ones. For example, Newton's Law of Inertia states that if no forces are acting upon a body, then its velocity remains the same. If a body were to change its velocity, it would follow that there were forces acting upon it. The Newtonian normal condition is that no forces are acting, and the Newtonian natural motion of a body is a motion with constant velocity. (The case of a zero velocity is a special case, so that remaining at rest is a special case of natural motion.)

This choice of natural motion and normalcy seems obvious and is due no doubt to their familiarity. Descartes and Newton, for example, shared the same notion of a natural motion, that is, nonacceleration. But Descartes used a different condition of normalcy. According to him, bodies move in straight lines with constant velocity unless they collide with other bodies. Evidently, he regarded the normal condition as one free of collision. Both Aristotle and Kepler differed from Newton over the normalcy condition. Kepler apparently believed it was normal for bodies to be at rest no matter where in the universe they might be, while Aristotle believed that there were distinct conditions of normalcy for distinct elements. It was normal only for earthly bodies to be at the center of the universe (which coincides with the center of the earth) where they would remain at rest.

We have described the Law of Inertia as a statement of the

normal condition, N, under which natural motion, M, of a body takes place. The statement has the form, in short: If N, then M. From such a statement it follows that non-M motions of bodies are cases of a non-N condition. But many scientists have expressed stronger demands upon the Law of Inertia, which emphasize its importance for theories of motion. Not only does a nonnatural motion occur whenever conditions are nonnormal, but these non-normal motions occur *because* there has been a departure from normalcy. Thus, Galileo believed that all terrestrial bodies move in a circle around the center of the earth, with a uniform speed, if conditions remain normal. He argued that it is change in the distance between a body and the source of gravity (the center of the earth) that causes changes in the body's velocity. In contrast to this circular inertia of Galileo, Newton proposed that it is the linear velocity which is preserved under normal conditions. All changes in this velocity, that is, all accelerations, are caused by nonnormal conditions, that is, by forces acting upon the body. Indeed, "force" (impressed force) for Newton had the sense of an external cause which changes a body's motion.

Both Galileo and Newton had a causal view of the inertia of bodies. They believed that two conditions hold true: (1) under normal conditions natural motion takes place, and (2) all non-natural motions are to be *explained* with the help of a theory of motion T and information about the kind of departure, from standard normal conditions, which takes place. If these two conditions are satisfied, the law "if N, then M" is *inertial with respect to* the theory T. Newton's First Law of Motion was regarded as a law of inertia with respect to the remainder of his theory of motion. It was believed that all deviations from natural motion (that is, all cases of accelerated motion) could be explained with the aid of his Second Law of Motion ("the total force on a body is equal to the product of its mass and its acceleration"), the Third Law of Motion ("the action of a force and the reactions to it are equal and opposite"), and his Law of Universal Gravitation.

If a statement of inertia, L, is inertial with respect to a theory of motion, T, and both L and T are believed to be true, then the motions of bodies receive a natural, satisfying systematization. Upon further empirical investigations, what could go wrong? Either L, T, or perhaps both statements might upon further evidence prove to be false.

If the "law" L proves, counter to our beliefs, to be false, there is the possibility of a modification of the conditions of "normalcy," "natural motion," or both to obtain a law L' which is inertial with respect to the theory T. This seems to have happened in the transition from the Galilean Law of Circular Inertia to that of Newton, where there was a shift from circular to rectilinear uniform motion as the natural motion of a body.

Upon further investigation, the theory T might prove to be false. Nevertheless one might discover a new theory T' so that L would be a law of inertia with respect to the new theory. Something like this seems to have happened when it was realized that Descartes' theory of motion was almost entirely incorrect, yet the Law of Inertia enunciated by him was believed sound. An improved theory of the collision of bodies along Cartesian lines was given by Christian Huyghens, John Wallis, and Sir Christopher Wren, and the Law of Inertia was believed to be inertial with respect to the improved theory.

There are cases where both the theory T and the law L are false. For example, according to Einstein, the classical Newtonian theory of motion is false when strictly taken. Relativistic mechanics yields a new theory of motion T' and a new law L' which is inertial relative to T'. The two laws, classical and relativistic, are different and the classical one is believed to be false. Each describes normal conditions and the natural motions of bodies, using radically different concepts of space and time. Yet, despite their diversity, each was believed in turn to be the Law of Inertia. Ernst Mach's work on the foundation of mechanics provided yet another example of a statement very different from Newton's which nevertheless was believed to be the Law of Inertia. According to Mach, all bodies that do not have forces acting upon them will not accelerate with respect to the fixed stars. A comparison of Mach's formulation of the Law of Inertia with Newton's shows that the chief difference lies in the use of the fixed stars, rather than Absolute Space, as a frame of reference.

Despite these changes, some of which are quite radical, these modifications are changes *in* the Law of Inertia. The Law of Inertia is understood to be something more than a statement which relates natural motions to normal conditions; it must also be inertial with respect to a theory of motion T. It is in virtue of this *relation* to a theory that two markedly different laws L and

L' can nonetheless be reckoned as laws of inertia. For example, suppose L' says that all free bodies move on geodesic paths, where a path is geodesic if and only if it is the shortest path between any two points lying on it. L' is inertial with respect to the General Theory of Relativity, which purports to explain all deviations from geodesic motion, even though L' is not equivalent to Newtonian inertia.

We ask how it is possible that so many different statements have each in turn been regarded as the Law of Inertia. The answer we have given raises new questions. A "law" L is the Law of Inertia if two conditions are satisfied: It states what natural motions take place under normal conditions, and secondly (the causal condition) L is inertial with respect to a theory T. The second condition is satisfied in each of the cases mentioned.

Is the notion of a normal condition and natural motion so wide that motions which are geodesic or circular or even rectilinear can count as natural motions? If so, then the only condition which has any bite to it is the one which requires L to be inertial with respect to a theory of motion T. But how restrictive is this requirement unless there is some understanding of "normal conditions" and "natural motion?" One restriction is that the conditions of normalcy and naturalness ought to be compatible with the theory T. For example, if the theory T requires that all motions be continuous, then it cannot be a natural motion for a body to be at two distinct places at two different times but at no intermediate place during that interval. Nor for that matter would "being under pressure of 10^8 pounds per square inch" be a normal condition if the theory of motion implies that bodies cannot exist under that quantity of pressure. Therefore, the theory T does impose some restrictions upon what is to count as normal and natural: these conditions must be compatible with T. Compatibility by itself is a very weak requirement, so weak that we may still feel that the statement "Under normal conditions, bodies will execute their natural motions," is contentless.

There is another condition which prevents the Law of Inertia from degenerating into an empty statement. The forces or causes which act upon bodies have somehow arisen *because* of the presence and possibly the motion of other bodies. Exactly how these other bodies give rise to the relevant forces is a matter for the theory to explain. Descartes believed that all changes from a straight path (with constant speed) were deviant and were caused by *collision* with other bodies. Newton, on the other hand, believed that the gravitational attraction of other bodies could cause certain noninertial motion. But he was extremely reluctant to say that these other bodies caused the gravitational effect to take place, since he had no acceptable theory, which would explain how the gravitational action of those other bodies was produced. Einstein believed that the motion of a body depends, according to his General Theory of Relativity, upon the distribution of matter throughout the universe. Hence, ultimately, the deviations from natural motion could be explained by invoking the distribution and state of motion of the other bodies in the universe. Einstein certainly thought that his theory did do this. He believed that his theory satisfied Mach's demand that the explanation of the motion of a body should refer to the action of all other bodies on it. And Mach, as we have already seen, believed this to be the major intuition behind Newton's laws of motion, however badly they were expressed.

If we believe that departures from natural motion are caused *somehow* by the presence and state of motion of other bodies, then we can conclude that all bodies should move naturally when no other bodies are present (provided, of course, as we have already mentioned, the condition is compatible with the theory of motion). Therefore, whatever else normal conditions may cover, they must include the case of isolated bodies.

Thus, certain conditions cannot be considered normal ones. Imagine that whenever exactly five bodies are present, their motions are of a natural kind. Suppose, too, that we possess a theory of motion T_5 which can explain all cases of nonnatural motion by taking into account the special ways in which there is divergence from the five-body case. These conditions alone would not guarantee that the five-body situation is a normal one. On the contrary, despite the achievements of the theory, no case of an isolated body is a five-body case, so that the presence of exactly five bodies cannot be a normal condition for natural motion.

In summary, in order for a certain statement to be a law of inertia, several conditions have to be satisfied:

1. Conditions of normalcy and naturalness are provided so that the law states that all bodies move naturally when normal conditions prevail.

2. There exists a theory of motion which can show how all deviations from natural motion arise from deviations from normal conditions (that is, the law is inertial with respect to the theory).

3. The normal condition must include the case of bodies in isolation, or something akin to this.

4. The conditions of normalcy and naturalness must be compatible with the theory of motion.

However, even if a law is inertial with respect to a theory, it does not follow that the theory explains the law. The theory explains how a nonnatural motion comes about through a deviation from normalcy. But it does not explain why there should be natural motion whenever things are normal. This is a delicate, logical point, but it is an important one. If a theory of motion does not explain the Law of Inertia, then there is a point to finding a theory which does explain it.

In recent times some very exciting attempts have been made to explain the Law of Inertia. We have included a sketch of the proposed explanation of the Law of Inertia by D. Sciama, who has pioneered in this area. The selections of this part of the book give a profile of the range and diversity of the specific conditions and theories, each offered as *the* Law of Inertia. The laws in each case are different, but the Law of Inertia has retained an invariant sense in that the relations between the conditions of normalcy, naturalness, and the relevant theory of motion have been preserved despite the changes which took place in any or all these factors. Each of the specific so-called laws of inertia are attempts to "fill in" or specify those factors of *the* Law of Inertia. Each time, it seemed that the proper theory and the normal and natural conditions had at last been found. The achievements of Kepler, Galileo, Descartes, Newton, and others included in this volume represent different versions of the Law of Inertia. They are different from each other, but they represent changes *in* the Law of Inertia rather than changes *of* the Law of Inertia. They represent a cumulative, focused sequence of investigations into the laws of motion.

Space

PLATO, from *The Timaeus*

Plato (*c*. 427–*c*. 347 B.C.) believed that all knowledge had to be a knowledge of things or objects which are real, and the reality of an object lay in its being eternal, unchanging and incorruptible. If we supplement this criterion of knowledge, with the view that all the objects known to us through our senses are changing and subject to decay, we obtain a central theme of Plato's thought: there is no true knowledge about objects which can be experienced. Despite this official view, Plato also believed that it was important to give some account of change, or becoming, and he found that the concepts of space and time are required to describe and explain change. If there are at least two bodies in the world, then there must be some third element, Space, which contains them. Space, according to Plato, is a kind of receptacle which contains all bodies. Although it has no visible qualities of its own, and there is no direct sensory evidence of its existence, nevertheless it does have a causal efficacy. The receptacle shakes the bodies in it from time to time, causing, he thought, similar bodies to cluster together. Plato's view bears a strong resemblance to Newton's, which also stated, as we shall see, that (Absolute) Space was a thing, not a body, which served as a container for bodies, and, in addition, had causal efficacy.

Tim. All men, Socrates, who have any degree of right feeling, at the beginning of every enterprise, whether small or great, always call upon God. And we, too, who are going to discourse of the nature of the universe, how created or how existing without creation, if we be not altogether out of our wits, must invoke the aid of Gods and Goddesses and pray that our words may be acceptable to them and consistent with themselves. Let this, then, be

Plato, *Timaeus*, tr. B. Jowett.

our invocation of the Gods, to which I add an exhortation of myself to speak in such manner as will be most intelligible to you, and will most accord with my own intent.

First then, in my judgment, we must make a distinction and ask, What is that which always is and has no becoming; and what is that which is always becoming and never is? That which is apprehended by intelligence and reason is always in the same state; but that which is conceived by opinion with the help of sensation and without reason, is always in a process of becoming and perishing and never really is. Now everything that becomes or is created must of necessity be created by some cause, for without a cause nothing can be created. The work of the creator, whenever he looks to the unchangeable and fashions the form and nature of his work after an unchangeable pattern, must necessarily be made fair and perfect; but when he looks to the created only, and uses a created pattern, it is not fair or perfect. Was the heaven then or the world, whether called by this or by any other more appropriate name—assuming the name, I am asking a question which has to be asked at the beginning of an enquiry about anything—was the world, I say, always in existence and without beginning? or created, and had it a beginning? Created, I reply, being visible and tangible and having a body, and therefore sensible; and all sensible things are apprehended by opinion and sense and are in a process of creation and created. Now that which is created must, as we affirm, of necessity be created by a cause. But the father and maker of all this universe is past finding out; and even if we found him, to tell of him to all men would be impossible. And there is still a question to be asked about him: Which of the patterns had the artificer in view when he made the world,—the pattern of the unchangeable, or of that which is created? If the world be indeed fair and the artificer good, it is manifest that he must have looked to that which is eternal; but if what cannot be said without blasphemy is true, then to the created pattern. Every one will see that he must have looked to the eternal; for the world is the fairest of creations and he is the best of causes. And having been created in this way, the world has been framed in the likeness of that which is apprehended by reason and mind and is unchangeable, and must therefore of necessity, if this is admitted, be a copy of something. Now it is all-important that the beginning of everything should be according

to nature. And in speaking of the copy and the original we may
assume that words are akin to the matter which they describe;
when they relate to the lasting and permanent and intelligible,
they ought to be lasting and unalterable, and, as far as their
nature allows, irrefutable and immovable—nothing less. But when
they express only the copy or likeness and not the eternal things
themselves, they need only be likely and analogous to the real
words. As being is to becoming, so is truth to belief. If then,
Socrates, amid the many opinions about the gods and the generation of the universe, we are not able to give notions which are
altogether and in every respect exact and consistent with one
another, do not be surprised. Enough, if we adduce probabilities
as likely as any others; for we must remember that I who am the
speaker, and you who are the judges, are only mortal men, and we
ought to accept the tale which is probable and enquire no further.

Soc. Excellent, Timaeus; and we will do precisely as you bid us.
The prelude is charming, and is already accepted by us—may we
beg of you to proceed to the strain?

.   .   .   .   .

Tim. This new beginning of our discussion of the universe requires a fuller division than the former; for then we made two
classes, now a third must be revealed. The two sufficed for the
former discussion: one, which we assumed, was a pattern intelligible and always the same; and the second was only the imitation of
the pattern, generated and visible. There is also a third kind which
we did not distinguish at the time, conceiving that the two would
be enough. But now the argument seems to require that we should
set forth in words another kind, which is difficult of explanation
and dimly seen. What nature are we to attribute to this new kind
of being? We reply, that it is the receptacle, and in a manner the
nurse, of all generation. I have spoken the truth; but I must express myself in clearer language, and this will be an arduous task
for many reasons, and in particular because I must first raise questions concerning fire and the other elements, and determine what
each of them is; for to say, with any probability or certitude, which
of them should be called water rather than fire, and which should
be called any of them rather than all or some one of them, is a
difficult matter. How, then, shall we settle this point, and what
questions about the elements may be fairly raised?

In the first place, we see that what we just now called water, by condensation, I suppose, becomes stone and earth; and this same element, when melted and dispersed, passes into vapour and air. Air, again, when inflamed, becomes fire; and again fire, when condensed and extinguished, passes once more into the form of air; and once more, air, when collected and condensed, produces cloud and mist; and from these, when still more compressed, comes flowing water, and from water comes earth and stones once more; and thus generation appears to be transmitted from one to the other in a circle. Thus, then, as the several elements never present themselves in the same form, how can any one have the assurance to assert positively that any of them, whatever it may be, is one thing rather than another? No one can. But much the safest plan is to speak of them as follows:—Anything which we see to be continually changing, as, for example, fire, we must not call 'this' or 'that,' but rather say that it is 'of such a nature;' nor let us speak of water as 'this,' but always as 'such;' nor must we imply that there is any stability in any of those things which we indicate by the use of the words 'this' and 'that,' supposing ourselves to signify something thereby; for they are too volatile to be detained in any such expressions as 'this,' or 'that,' or 'relative to this,' or any other mode of speaking which represents them as permanent. We ought not to apply 'this' to any of them, but rather the word 'such;' which expresses the similar principle circulating in each and all of them; for example, that should be called 'fire' which is of such a nature always, and so of everything that has generation. That in which the elements severally grow up, and appear, and decay, is alone to be called by the name 'this' or 'that;' but that which is of a certain nature, hot or white, or anything which admits of opposite qualities, and all things that are compounded of them, ought not to be so denominated. Let me make another attempt to explain my meaning more clearly. Suppose a person to make all kinds of figures of gold and to be always transmuting one form into all the rest;—somebody points to one of them and asks what it is. By far the safest and truest answer is, That is gold; and not to call the triangle or any other figures which are formed in the gold 'these,' as though they had existence, since they are in process of change while he is making the assertion; but if the questioner be willing to take the safe and indefinite expression, 'such,' we should be satisfied. And the same argument applies to the universal nature which receives all bodies—that must be always called the same; for, while receiving all things, she never departs at all from her own nature, and never in any way, or at any time, assumes a form like that of any of the things which enter into her; she is the natural recipient of all impressions, and is stirred and informed by them, and appears different from time to time by reason of them. But the forms which enter into and go out of her are the likenesses of real existences modelled after their patterns in a wonderful and inexplicable manner, which we will hereafter investigate. For the present we have only to conceive of three natures: first, that which is in process of generation; secondly, that in which the generation takes place; and thirdly, that of which the thing generated is a resemblance. And we may liken the receiving principle to a mother, and the source or spring to a father, and the intermediate nature to a child; and may remark further, that if the model is to take every variety of form, then the matter in which the model is fashioned will not be duly prepared, unless it is formless, and free from the impress of any of those shapes which it is hereafter to receive from without. For if the matter were like any of the supervening forms, then whenever any opposite or entirely different nature was stamped upon its surface, it would take the impression badly, because it would intrude its own shape. Wherefore, that which is to receive all forms should have no form; as in making perfumes they first contrive that the liquid substance which is to receive the scent shall be as inodorous as possible; or as those who wish to impress figures on soft substances do not allow any previous impression to remain, but begin by making the surface as even and smooth as possible. In the same way that which is to receive perpetually and through its whole extent the resemblances of all eternal beings ought to be devoid of any particular form. Wherefore, the mother and receptacle of all created and visible and in any way sensible things, is not to be termed earth, or air, or fire, or water, or any of their compounds, or any of the elements from which these are derived, but is an invisible and formless being which receives all things and in some mysterious way partakes of the intelligible, and is most incomprehensible. In saying this we shall not be far wrong; as far, however, as we can attain to a knowledge of her from the previous considerations, we may truly say that fire is that part of her nature which from time to time is inflamed, and water that which is moistened, and that the mother substance becomes earth and air, in so far as she receives the impressions of them.

Let us consider this question more precisely. Is there any self-existent fire? and do all those things which we call self-existent exist? or are only those things which we see, or in some way perceive through the bodily organs, truly existent, and nothing whatever besides them? And is all that which we call an intelligible essence nothing at all, and only a name? Here is a question which we must not leave unexamined or undetermined, nor must we affirm too confidently that there can be no decision; neither must we interpolate in our present long discourse a digression equally long, but if it is possible to set forth a great principle in a few words, that is just what we want.

Thus I state my view:—If mind and true opinion are two distinct classes, then I say that there certainly are these self-existent ideas unperceived by sense, and apprehended only by the mind; if, however, as some say, true opinion differs in no respect from mind, then everything that we perceive through the body is to be regarded as most real and certain. But we must affirm them to be distinct, for they have a distinct origin and are of a different nature; the one is implanted in us by instruction, the other by persuasion; the one is always accompanied by true reason, the other is without reason; the one cannot be overcome by persuasion, but the other can: and lastly, every man may be said to share in true opinion, but mind is the attribute of the gods and of very few men. Wherefore also we must acknowledge that there is one kind of being which is always the same, uncreated and indestructible, never receiving anything into itself from without, nor itself going out to any other, but invisible and imperceptible by any sense, and of which the contemplation is granted to intelligence only. And there is another nature of the same name with it, and like to it, perceived by sense, created, always in motion, becoming in place and again vanishing out of place, which is apprehended by opinion and sense. And there is a third nature, which is space, and is eternal, and admits not of destruction and provides a home for all created things, and is apprehended without the help of sense, by a kind of spurious reason, and is hardly real; which we beholding as in a dream, say of all existence that it must of necessity be in some place and occupy a space, but that what is neither in heaven nor in earth has no existence. Of these and other things of the same kind, relating to the true and waking reality of nature, we have only this dreamlike sense, and we are unable to cast off sleep and determine the truth about them. For an image, since the reality, after which it is modelled, does not belong to it, and it exists ever as the fleeting shadow of some other, must be inferred to be in another [i.e. in space], grasping existence in some way or other, or it could not be at all. But true and exact reason, vindicating the nature of true being, maintains that while two things [i.e. the image and space] are different they cannot exist one of them in the other and so be one and also two at the same time.

Thus have I concisely given the result of my thoughts; and my verdict is that being and space and generation, these three, existed in their three ways before the heaven; and that the nurse of generation, moistened by water and inflamed by fire, and receiving the forms of earth and air, and experiencing all the affections which accompany these, presented a strange variety of appearances; and being full of powers which were neither similar nor equally balanced, was never in any part in a state of equipoise, but swaying unevenly hither and thither, was shaken by them, and by its motion again shook them; and the elements when moved were separated and carried continually, some one way, some another; as, when grain is shaken and winnowed by fans and other instruments used in the threshing of corn, the close and heavy particles are borne away and settle in one direction, and the loose and light particles in another. In this manner, the four kinds or elements were then shaken by the receiving vessel, which, moving like a winnowing machine, scattered far away from one another the elements most unlike, and forced the most similar elements into close contact. Wherefore also the various elements had different places before they were arranged so as to form the universe. At first, they were all without reason and measure. But when the world began to get into order, fire and water and earth and air had only certain faint traces of themselves, and were altogether such as everything might be expected to be in the absence of God; this, I say, was their nature at that time, and God fashioned them by form and number. Let it be consistently maintained by us in all that we say that God made them as far as possible the fairest and best, out of things which were not fair and good. And now I will endeavour to show you

the disposition and generation of them by an unaccustomed argument, which I am compelled to use; but I believe that you will
be able to follow me, for your education has made you familiar
with the methods of science.

.   .   .   .   .

Tim. Now all unmixed and primary bodies are produced by
such causes as these. As to the subordinate species which are included in the greater kinds, they are to be attributed to the varie-
ties in the structure of the two original triangles. For either structure did not originally produce the triangle of one size only, but
some larger and some smaller, and there are as many sizes as there
are species of the four elements. Hence when they are mingled
with themselves and with one another there is an endless variety
of them, which those who would arrive at the probable truth of
nature ought duly to consider.

Unless a person comes to an understanding about the nature
and conditions of rest and motion, he will meet with many difficulties in the discussion which follows. Something has been said
of this matter already, and something more remains to be said,
which is, that motion never exists in what is uniform. For to
conceive that anything can be moved without a mover is hard
or indeed impossible, and equally impossible to conceive that
there can be a mover unless there be something which can be
moved;—motion cannot exist where either of these are wanting,
and for these to be uniform is impossible; wherefore we must
assign rest to uniformity and motion to the want of uniformity.

Time

PLATO, from *The Timaeus*

When the father and creator saw the creature which he had made moving and living, the created image of the eternal gods, he rejoiced, and in his joy determined to make the copy still more like the original; and as this was eternal, he sought to make the universe eternal, so far as might be. Now the nature of the ideal being was everlasting, but to bestow this attribute in its fulness upon a creature was impossible. Wherefore he resolved to have a moving image of eternity, and when he set in order the heaven, he made this image eternal but moving according to number, while eternity itself rests in unity; and this image we call time. For there were no days and nights and months and years before the heaven was created, but when he constructed the heaven he created them also. They are all parts of time, and the past and future are created species of time, which we unconsciously but wrongly transfer to the eternal essence; for we say that he 'was,' he 'is,' he 'will be,' but the truth is that 'is' alone is properly attributed to him, and that 'was' and 'will be' are only to be spoken of becoming in time, for they are motions, but that which is immovably the same cannot become older or younger by time, nor ever did or has become, or hereafter will be, older or younger, nor is subject at all to any of those states which affect moving and sensible things and of which generation is the cause. These are the forms of time, which imitates eternity and revolves according to a law of number. Moreover, when we say that what has become *is* become and what becomes *is* becoming, and that what will become *is* about to become and that the non-existent *is* non-existent,—all these are inaccurate modes of expression. But perhaps this whole subject will be more suitably discussed on some other occasion.

Time, then, and the heaven came into being at the same in-

Plato, *Timaeus,* tr. B. Jowett.

stant in order that, having been created together, if ever there was to be a dissolution of them, they might be dissolved together. It was framed after the pattern of the eternal nature, that it might resemble this as far as was possible; for the pattern exists from eternity, and the created heaven has been, and is, and will be, in all time. Such was the mind and thought of God in the creation of time. The sun and moon and five other stars, which are called the planets, were created by him in order to distinguish and preserve the numbers of time; and when he had made their several bodies, he placed them in the orbits in which the circle of the other was revolving (cp. 36 D),—in seven orbits seven stars. First, there was the moon in the orbit nearest the earth, and next the sun, in the second orbit above the earth; then came the morning star and the star sacred to Hermes, moving in orbits which have an equal swiftness with the sun, but in an opposite direction; and this is the reason why the sun and Hermes and Lucifer overtake and are overtaken by each other. To enumerate the places which he assigned to the other stars, and to give all the reasons why he assigned them, although a secondary matter, would give more trouble than the primary. These things at some future time, when we are at leisure, may have the consideration which they deserve, but not at present.

Natural and Forced Motions

ARISTOTLE, from *The Physics*

Aristotle (*b*. 384–322 B.C.) probably wrote parts of the *Physics*
not long after leaving Plato's Academy upon the latter's death
in 348–347 B.C. It contains Zeno's celebrated *Paradoxes of Motion,* consisting of four arguments, each designed to show that
the common descriptions of motion lead to consequences which
are inconsistent with widely shared beliefs. Zeno thought that
he had thereby shown the impossibility of change or motion.

Aristotle, however, believed that changes do take place, and his
analysis of the concept of motion was to dominate, if not strongly
influence, the analyses of scientists for almost two thousand years.
Central to his analyses were the theses that all actual motion requires a mover as well as a moved object; that there is a prime
mover or motor which is motionless, dimensionless, eternal, and
the cause of all motion in the world; that motion is eternal and
imperishable and that motions can be divided into two kinds:
natural and forced. Self-moved things move naturally, according
to him. But the converse is not true. The reason is that light
bodies move upward naturally, but they do not move upward
themselves. If they did, they would be self-moved, and this would
mean that they were alive! Thus, the major task in the *Physics*
was to specify the agent which causes bodies to move. The distinction between natural and forced motions became a prominent way
of organizing the problems of physics, even for those physicists
who wanted to argue against Aristotle (see the Copernicus and
Galileo selections) and it played a key role in the development
of the Law of Inertia (see the Galileo reading). The concepts of
place and time were developed with close attention to Plato's conclusions. The place of a body exists. It is not a body, but it does
have causal power, since certain kinds of bodies tend toward certain fixed places. And the places of bodies serve to distinguish

Reprinted by permission of the publishers and the Loeb Classical Library,
from Philip Wicksteed and F. M. Cornford, translators, Aristotle, *The Physics,*
Vols. I & II, Cambridge, Mass.: Harvard University Press.

and to locate them. Time is not a motion (this suggests a contrast with Plato's view that time is the moving image of Eternity); rather, it is a numerical measure of motion.

Zeno's Paradoxes of Motion

The fallacy in Zeno's argument is now obvious; for he says that since a thing is at rest when it has not shifted in any degree out of a place equal to its own dimensions, and since at any given instant during the whole of its supposed motion the supposed moving thing is in the place it occupies at that instant, the arrow is not moving at *any time* during its flight. But this is a false conclusion; for time is not made up of atomic 'nows,' any more than any other magnitude is made of up atomic elements.[1]

Of Zeno's arguments about motion, there are four which give trouble to those who try to solve the problems they raise. The first is the one which declares movement to be impossible because, however near the mobile is to any given point, it will always have to cover the half, and then the half of that, and so on without limit before it gets there. And this we have already taken to pieces.[2]

The second is what is known as 'the Achilles,' which purports to show that the slowest will never be overtaken in its course by the swiftest, inasmuch as, reckoning from any given instant, the pursuer, before he can catch the pursued, must reach the point from which the pursued started at that instant, and so the slower will always be some distance in advance of the swifter. But this argument is the same as the former one which depends on bisection, with the difference that the division of the magnitudes we successively take is not a division into halves (but according to any ratio we like to assume between the two speeds). The conclusion of the argument is that the slower cannot be overtaken by the swifter, but it is reached by following the same lines as the 'bisection' argument of the first thesis; for the reason why neither supposed process lands us at the limit, is that the method of division is expressly so designed as not to get us there, only in this second thesis a declamatory intensification is introduced by representing the swiftest racer as unable to overtake the slowest. The solution then must be identical in both cases, and the claim that the thing that is ahead is not overtaken is false. It is not over-taken *while* it is ahead, but none the less *is* overtaken if Zeno will allow it to traverse to the end its finite distance.[3] So much for these two theses.

The third thesis is the one just mentioned, namely that the arrow is stationary while on its flight. The demonstration rests on the assumption that time is made up of 'nows,' and if this be not granted the inference fails.

The fourth thesis supposes a number of objects all equal with each other in dimensions, forming two equal trains and arranged so that one train stretches from one end of a racecourse to the middle of it, and the other from the middle to the other end. Then if you let the two trains, moving in opposite directions but at the same rate, pass each other, Zeno undertakes to show that half of the time they occupy in passing each other is equal to the whole of it. The fallacy lies in his assuming that a moving object takes an equal time in passing another object equal in dimen-sions to itself, whether that other object is stationary or in mo-tion; which assumption is false. For this is his demonstration. Let there be a number of objects AAAA, equal in number and bulk to those that compose the two trains but stationary in the middle of the stadium. Then let the objects BBBB, in number and dimension equal to the A's, form one of the trains stretching from the middle of the A's in one direction; and from the inner end of the B's let CCCC stretch in the opposite direction, being equal in number, dimension, and rate of movement to the B's.

Then when they cross, the first B and the first C will simultane-ously reach the extreme A's in contrary directions.

Now during this process the first C has passed all the B's, whereas the first B has only passed half the A's, and therefore only taken half the time; for it takes an equal time (the minimal time) for the C to pass one B as for the B to pass one A. But during this

same half-time the first B has also passed all the C's[4] (though the
first B takes as long, says Zeno, to pass a C as an A) because
measured by their progress through the A's the B's and C's have
had the same time in which to cross each other.[5] Such is his
argument, but the result depends on the fallacy above mentioned.[6]

Nor need we be troubled by any attack on the possibility of
change based on the axiom that a thing 'must either be or not
be' but cannot 'both be and not be' this or that at the same
time. For, it is argued, if a thing is changing, for instance, from
being not-white to being white and is on its way from one to the
other, you can truly assert at the same time that it is neither
white nor not white. But this is not true, for we sometimes call
a thing 'white' even if it is not entirely white, and we sometimes
call a thing 'not-white' even if there is some trace of white in it;
we speak of it according to its prevailing condition or the conditions of its most significant parts or aspects. For to say that a
thing is not in a certain condition 'at all' and to say that it is
not 'altogether,' in it are two different things. And so, too, in
the case of being or not being or any other pair of contradictory
opposites. For during the whole process of changing it must be
prevailingly one or the other and can never be exclusively either.[7]

Again it is said that a rotating circle or sphere or anything
else that moves within its own dimensions is stationary because
in itself and all its parts it will remain in the same place for the
given time: so it will be in motion and at rest at the same time.
But in the first place its parts are not in the same place during
any space of time, and in the second place the whole is also con
tinuously changing to a different (rotational) position;[8] for the
circumference measured round from A to A again is not identical
with the circumference as measured from B to B or from C to C
or any other point, except by accidental concomitance, as the
cultivated person is a man. Thus one circumference is ever succeeding another and it will never be at rest. So, too, with the
sphere, and any other body that moves within fixed dimensions.
(*Physics*, VI, ix)

·   ·   ·   ·   ·

Motion takes place, then, when there exists a mobile and a
potential motor and they are so approximated that the one really
is able to act and the other to be acted on. So if they had been

there eternally but without motion, it must obviously have been because they were not in such relations as to make them *actually able* to cause motion or to be moved. For motion to supervene, therefore, it must be necessary that one or the other should experience a change, for this must be so where we are dealing with any pair of related factors—for instance if A is now twice B and was not so before, either one or both must have changed. So there would have to be a change anterior to the supposed first change.

And besides this, how could there be any before or after at all if time were not, or time itself be if there were no motion? For surely if time is the numerical aspect of movement[9] or is itself a movement, it follows that, if there has always been time, there must always have been movement; and as to that it seems that, with a single exception, all thinkers agree that time never came into existence but was always there. It is thus that Democritus shows how impossible it is that everything can have had an origin —because time has not. (Plato alone assigns an origin to time, for he says it came into existence simultaneously with the universe[10] and he assigns an origin to that.) Well then, if it is impossible for time to exist or to be conceived without the 'present now,' and if this 'now' is a kind of midmostness, which combines beginning and end—the beginning, to wit, of future and the end of past time—then there must always have been time; for however far back you go in time past, the extreme limit you take must be a certain 'now' (for in time there is nothing else to take except a 'now'), and since every now is an end as well as a beginning, it follows that time stretches from it in both directions. And if time, then motion, inasmuch as time is but an aspect or affection of motion.

The same line of argument further shows that movement is imperishable. For as we have seen that if we suppose movement to have had an origin we shall have to suppose that there was a change anterior to the first change, so also if we suppose it to cease we shall have to admit a change posterior to the last change. For what is movable does not cease to be movable because it is no longer being moved, nor does that which is capable of causing motion cease to have that capacity because it is not moving anything; for instance the combustible if it is not being burnt (for it may still be combustible all the same) or the potential agent of local shifting when it is shifting nothing. And so, if all the destructible were destroyed that would not destroy the destroying agent, which would remain for destruction in its turn, and when it was destroyed its destroyer would remain; and being destroyed is a kind of change.[11] And if all this is impossible, it is evident that movement is eternal, and is not something which now was and now was not. Indeed to assert the opposite is very like a contradiction in terms.[12] (*Physics*, VIII, i)

. . . . .

Of the proper subjects of motion some are moved by themselves and others by something not themselves, and some have a movement natural to themselves and others have a movement forced upon them which is not natural to them. Thus the self-moved has a natural motion.

The real difficulty then is narrowed down to those movements of things that are not self-moving which we have not yet dealt with; for in pronouncing some of the movements of things which are not self-moving to be contrary to their nature, we have by inference laid down that the rest are natural; and it is here that we come to grips with the real difficulty, viz. the question what is the agent of the natural movements of bodies heavy and light. For such bodies can be forced to move in directions opposite to those natural to them; but whereas it is obvious that light things go up and heavy ones down 'by nature,' we have not yet arrived at any clear conception as to what is the agent of this 'natural' movement, as we have done in the case of the enforced and unnatural movements.

For we cannot say that such bodies, when moving naturally, 'move themselves,' for this is proper to animals that have life, and if light and heavy bodies moved themselves up and down they would be able to stop themselves also—I mean that if an animal can make itself march it can also make itself stop marching—so that if fire makes itself move upwards it should be able to make itself move downwards also. If they moved themselves there would be no sense in saying that they could only move in one direction.[13] Again what can be meant by a continuous and homogeneous body 'moving itself'? For in so far as it is one and continuous (otherwise than by contact)[14] it cannot be affected by itself; it is only if it can be analysed into parts or factors that it can be self-affected, by one of its elements being the agent and another the patient. Thus[15] in the movement of such bodies there is no single constituent which is at once the primary agent and the primary patient (nor can there be, since each constituent is homogeneous in itself). And the argument applies to all properly 'continuous' bodies. So in every case the mover and the moved must be distinguished, just as we see that they are when a living thing as agent moves a lifeless thing as patient. These continuous bodies are, in fact, always moved by something else; what this something is would become clear if we were to distinguish the causes involved.

The above-mentioned distinctions can be drawn in the case of the agents of motion: some of them are capable of causing motion unnaturally (a lever, for instance, is not by nature capable of moving a load),[16] others naturally; thus a body that is actually hot is capable of causing a change in one that is potentially hot; and so with other kinds of change. And it is the same with what is capable of suffering change: the natural subject of change is that which is potentially of a certain quality or quantity or in a certain place, when it contains the principle of the modification in question in itself and not accidentally—'not accidentally,' because a thing that comes to have a certain quality may also grow to a certain size at the same time, but the change of size is incidental to the change of quality and is not an essential property of the thing *qua* capable of qualitative change. So, then, when fire and earth are moved by some agent, whereas the motion is forcible when it is contrary to their nature, it is natural when they actually engage in their proper movements, the potentiality for which was already inherent in them. (*Physics*, VIII, iv)

.   .   .   .   .

We have now proved that an unlimited force cannot reside in a limited magnitude, and also that the force residing in an unlimited magnitude cannot itself be limited.

But before discussing rotating bodies it will be well to examine a certain question concerning bodies in locomotion. If everything that is in motion is being moved by something, how comes it that certain things, missiles for example, that are not self-moving nevertheless continue their motion without a break when no longer in contact with the agent that gave them motion? Even if that agent at the same time that he puts the missile in motion also sets something else (say air) in motion, which something when itself in motion has power to move other things, still when the prime agent has ceased to be in contact with this secondary agent and has therefore ceased to be moving it, it must be just as impossible for it as for the missile to be in motion: missile and secondary agent must all be in motion simultaneously, and must have ceased to be in motion the instant the prime mover ceases to move them; and this holds good even if the prime agent is like the magnet, which has power to confer upon the iron bar it moves the power of moving another iron bar.[17] We are forced, therefore, to suppose that the prime mover conveys to the air (or water, or other such intermediary as is naturally capable both of moving and conveying motion) a power of conveying motion, but that this power is not exhausted when the intermediary ceases to be moved itself. Thus the intermediary will cease to be moved itself as soon as the prime mover ceases to move it, but will still be able to move something else. Thus this something else will be put in motion after the prime mover's action has ceased, and will itself continue the series. The end of it all will approach as the motive power conveyed to each successive secondary agent wanes, till at last there comes one which can only move its neighbor without being able to convey motive force to it. At this point the last active intermediary will cease to convey motion, the passive intermediary that has no active power will cease to be in motion, and the missile will come to a stand, at the same instant.

All these points being established, it is clear that the prime and motionless motor cannot have magnitude. For if it had magnitude it must either be limited or unlimited. Now we have shown already, in our discourse on Physics,[18] that there cannot be an unlimited magnitude; and now we have further proved that a limited body cannot exert unlimited force and that nothing can be kept in motion for an unlimited time by a limited agent. But the prime motor causes an everlasting motion and maintains it through time without limit. It is manifest, therefore, that this prime motor is not divisible, has no parts, and is not dimensional. (*Physics,* VIII, x.)

.   .   .   .   .

The Natural Philosopher has to ask the same questions about 'place' as about the 'unlimited'; namely, whether such a thing exists at all, and (if so) after what fashion it exists, and how we are to define it.

It is generally assumed that whatever exists, exists 'some-where'[19] (that is to say, 'in some place'), in contrast to things which 'are nowhere' because they are non-existent,—so that the obvious answer to the question "where is the goat-stag or the sphinx?" is "nowhere." And again the primary and most general case of 'passage' or transitional change from 'this' to 'that' is the case of local change from this to that 'place.'

But we encounter many difficulties when we attempt to say what exactly the 'place' of a thing is. For according to the data from which we start we seem to reach different and inconsistent conclusions. Nor have my precursors laid anything down, or even formulated any problems, on this subject.

To begin with, then, the phenomenon of 'replacement' seems at once to prove the independent existence of the 'place' from which—as if from a vessel—water, for instance, has gone out, and into which air has come, and which some other body yet may occupy in its turn; for the place itself is thus revealed as some-ing different from each and all of its changing contents. For 'that wherein' air *is,* is identical with 'that wherein' water *was*; so that the 'place' or 'room' into which each substance came, or out of which it went, must all the time have been distinct from both of the substances alike.

Moreover the trends of the physical elements (fire, earth, and the rest) show not only that locality or position is a reality but also that it exerts an active influence; for fire and earth are borne, the one upwards and the other downwards, if unimpeded, each towards its own 'position,' and these terms—'up' and 'down' I mean, and the rest of the six dimensional directions—indicate subdivisions or distinct classes of positions or localities in general.

Now these terms—such as up and down and right and left, I mean—when thus applied to the trends of the elements are not merely relative to ourselves. For in this relative sense the terms have no constancy, but change their meaning according to our own position, as we turn this way or that; so that the same thing may be now to the right and now to the left, now above and now below, now in front and now behind; whereas in Nature each of these directions is distinct and stable independently of us. 'Up' or 'above' always indicates the 'whither' to which things buoyant tend; and so too 'down' or 'below' always indicates the 'whither' to which weighty and earthy matters tend, and does not change with circumstance; and this shows that 'above' and 'below' not only indicate definite and distinct localities, directions, and positions, but also produce distinct effects. The comparison of mathematical figures illustrates the point. For such figures occupy no real positions of their own, but nevertheless acquire a right and left with reference to us, thus showing that their positions are merely such as we mentally assign to them and are not intrinsically distinguished by anything in Nature.

Further, the thinkers who assert the existence of the 'void' agree with all others in recognizing the reality of 'place,' for the 'void' is supposed to be 'place without anything in it.'

One might well conclude from all this that there must be such a thing as 'place' independent of all bodies, and that all bodies cognizable by the senses occupy their several distinct places. And this would justify Hesiod in giving primacy to *Chaos* [=the 'Gape'] where he says: "First of all things was Chaos, and next broad-bosomed Earth"; since before there could be anything else 'room' must be provided for it to occupy. For he accepted the general opinion that everything must be somewhere and must have a place.

And if such a thing should really exist well might we contemplate it with wonder—capable as it must be of existing without anything else, whereas nothing else could exist without it, since 'place' is not destroyed when its contents vanish.

But then, if we grant that such a thing exists, the question as to *how* it exists and what it really is must give us pause. Is it some kind of corporeal bulk? Or has it some other mode of existence? For we must begin by determining to what category it belongs.[20]

Now a 'place,' as such, has the three dimensions of length, breadth, and depth, which determine the limits of all bodies; but it cannot itself be a body, for if a 'body' were in a 'place' and the place itself were a body, two bodies would coincide.

Another difficulty. If a body has 'place' and 'room,' the reasoning already employed would show that its surfaces and other limits must also have them. For 'where' the surfaces of the water were, 'there' the surfaces of the air now are. But we cannot distinguish between a point and its own position, and if there is no distinction here neither is there in any of the others, line, surface, bulk or capacity. So what would then become of the thesis that all bodies have places distinguishable from themselves?

What kind of thing then must we conceive 'a place' to be? Its properties forbid us to think of it either as being an element itself or as being compounded of elements—whether physical or conceptual. For it has size, which conceptual components could not give it; but it is not a body, which it would necessarily be if compounded of elements cognizable by sense.[21]

Again, how are we to suppose that place affects or determines things in any way? For it cannot be brought under any one of the four causal or essential determinants:—not as the 'material' of things, for nothing is composed of it; nor as their 'form' or constituent definition; nor as their contemplated 'end'; nor as setting them in motion, or otherwise changing them.[22]

And yet again: If it has an existence of its own, 'where' does it exist? For we cannot ignore Zeno's dilemma:[23] If everything that exists, exists in some 'place,' then if the place itself exists it too must have a place to exist in, and so on *ad infinitum*.

Further: If each body exactly occupies the place it is in, then reciprocally each place is exactly occupied by the body in it. But in that case what account are we to give of 'growing' things? It would seem that their places must also grow, to keep company with them, since they can never be less than the places they occupy, nor the places they occupy be greater than they are.

So, after all, we are forced by these perplexities not only to ask what a 'place' is, but also to reopen the question that appeared to be closed, and ask whether there is such a thing as 'place' at all.

It should now be possible to reach a clear conception of what the place of a thing really is, by gathering up such properties of 'place' as appear to have emerged securely as its characteristics, and proceeding as set forth in the sequel.

Well then, to begin with, we may safely assert—(i.) That the place of a thing is no part or factor of the thing itself, but is that which embraces it; (ii.) that the immediate or 'proper' place of a thing is neither smaller nor greater than the thing itself; (iii.) that the place where the thing is can be quitted by it, and is therefore separable from it; and lastly, (iv.) that any and every place implies and involves the correlatives of 'above' and 'below,' and that all the elemental substances have a natural tendency to move towards their own special places, or to rest in them when there—such movement being 'upward' or 'downward,' and such rest 'above' or 'below.'

Taking these as our data we may now pursue our investigation, and we just try so to conduct it (1) that a full account shall be given of the meaning and nature of 'place'; (2) that the problems we have encountered on our way shall find their solution in it; (3) that the characteristics of 'place' which we have noted shall reveal themselves as integral to its nature as we have defined it; and lastly (4) that the perplexities we have met shall be seen to rise naturally out of the facts, as explained. No more can be demanded of any solution than that it should satisfy all these conditions.

To begin with, then, we must recognize that no speculations as to place would ever have arisen had there been no such thing as movement, or *change* of place. Indeed, the chief reason of the persistent tendency to think of heaven itself as having a 'place' is that it is always moving.[24]

Now change of place may occur either by way of translation, or by way of increase and decrease, for in this case too what was formerly in such and such a place is now in a larger or a smaller one.

Again, of things which are in motion some are moved by the actualizing of their own inherent potentialities, and others only by being involved in the movement of something else in which they inhere; and of this latter class again some (being substantive) are *capable* of movement on their own account as well (such are limbs of the body, or a rivet in a ship), whereas others, like 'whiteness' or 'wisdom' (being attributive) are not so much as capable of moving otherwise than incidentally, for the only sense in which they move at all is that they inhere in something that moves.

Again,[25] when we say that a thing is 'in the universe,' as though that were its place, it is because it is in the air, which air is in the universe. And we say it is in the air not because it is in all of it, but in virtue of its being embraced by a certain surface of the air, which fits around it; for if the whole of the air were its place, then the place a thing occupied would not be of the same size as the thing itself, which equality of size we have accepted as one of our data.

Such then—the inner surface of the envelope, namely—is the immediate place of a thing.

·   ·   ·   ·   ·

The subject of inquiry next in succession is 'time.' It will be well to begin with the questions which general reflections suggest as to its existence or non-existence and its nature.

The following considerations might make one suspect either that there is really no such thing as time, or at least that it has only an equivocal and obscure existence.

(1) Some of it is past and no longer exists, and the rest is future and does not yet exist; and time, whether limitless or any given length of time we take, is entirely made up of the no-longer and not-yet; and how can we conceive of that which is composed of non-existents sharing in existence in any way?

(2) Moreover, if anything divisible exists, then, so long as it is in existence, either all its parts or some of them must exist. Now time is divisible into parts, and some of these were in the past and some will be in the future, but none of them exists. The present 'now' is not part of time at all, for a part measures the whole, and the whole must be made up of the parts, but we cannot say that time is made up of 'nows.'

(3) Nor is it easy to see whether the 'now' that appears to divide the past and the future (a) is always one and the same or (b) is perpetually different.

(b) For if it is perpetually different, and if no two sectional parts of time can exist at once (unless one includes the other, the longer the shorter), and if the 'now' that is not, but was, must have ceased to be at *some* time or other,[26] so also no two 'nows' can exist together, but the past 'now' must have perished before there was any other 'now.' Now it cannot have ceased to be when it was itself the 'now,' for that is just when it existed; but it is impossible that the past 'now' should have perished in any other 'now' but itself. For we must lay it down as an axiom that there can be no next 'now' to a given 'now,' any more than a next point to a given point.[27] So that if it did not perish in the next 'now,' but in some subsequent one, it would have been in existence coincidently with the countless 'nows' that lie between the 'now' in which it was and the subsequent 'now' in which we are supposing it to perish; which is impossible.

(a) But neither can it continuously persist in its identity. For nothing which is finite and divisible is bounded by a single limit, whether it be continuous in one dimension only or in more than one; but the 'now' is a time limit, and if we take any limited period of time, it must be determined by two limits, which cannot be identical. Again, if simultaneity in time, and not being before or after, means coinciding and being in the very 'now' wherein they coincide, then, if the before and the after were both in the persistently identical 'now' we are discussing, what happened ten thousand years ago would be simultaneous with what is happening to-day, and nothing would be before or after anything else.

Let this suffice as to the problems raised by considering the properties of time.

But what time really is and under what category it falls, is no more revealed by anything that has come down to us from earlier thinkers than it is by the considerations that have just been urged. For (a) some[28] have identified time with the revolution of the all-embracing heaven, and (b) some[29] with that heavenly sphere itself. But (a) a partial revolution is time just as much as a whole one is, but it is not just as much a revolution; for any finite portion of time is a portion of a revolution, but is not a revolution.[30] Moreover, if there were more universes than one, the re-entrant circumlation of each of them would be time, so that several different times would exist at once. And (b) as to those who declare the heavenly sphere itself to be time, their only reason was that all things are contained 'in the celestial sphere' and also occur 'in time,' which is too childish to be worth reducing to absurdities more obvious than itself.

Now the most obvious thing about time is that it strikes us as some kind of 'passing along' and changing; but if we follow this clue, we find that, when any particular thing changes or moves, the movement or change is in the moving or changing thing itself or occurs only where that thing is; whereas 'the passage of time' is current everywhere alike and is in relation with everything. And further, all changes may be faster or slower, but not so time; for fast and slow are defined by time, 'faster' being more change in less time, and 'slower' less in more. But time cannot measure time thus, as though it were a distance (like the space passed through in motion) or a qualitative modification, as in other kinds of change.[31] It is evident, therefore, that time is not identical with movement; nor, in this connexion, need we distinguish between movement and other kinds of change.

. . . . . .

On the other hand, time cannot be disconnected from change; for when we experience no changes of consciousness, or, if we do, are not aware of them, no time seems to have passed, any more than it did to the men in the fable who 'slept with the heroes'[32] in Sardinia, when they awoke; for under such circumstances we fit the former 'now' onto the later, making them one and the same and eliminating the interval between them, because we did not perceive it. So, just as there would be no time if there were no distinction between this 'now' and that 'now,' but it were always the same 'now'; in the same way there appears to be no time between two 'nows' when we fail to distinguish between them. Since, then, we are not aware of time when we do not distinguish any change (the mind appearing to abide in a single indivisible and undifferentiated state), whereas if we perceive and distinguish changes, then we say that time has elapsed, it is clear that time cannot be disconnected from motion and change.

Plainly, then, time is neither identical with movement nor capable of being separated from it.

In our attempt to find out what time is, therefore, we must start from the question, in what way it pertains to movement. For when we are aware of movement we are thereby aware of time, since, even if it were dark and we were conscious of no bodily sensations, but something were 'going on' in our minds, we should, from that very experience, recognize the passage of time. And conversely, whenever we recognize that there has been a lapse of time, we by that act recognize that something 'has been going on.' So time must either itself be movement, or if not, must pertain to movement; and since we have seen that it is not identical with movement, it must pertain to it in some way.

Well then, since anything that moves moves from a 'here' to a 'there,'[33] and magnitude as such is continuous, movement is dependent on magnitude; for it is because magnitude is continuous that movement is also, and because movement is continuous so is time; for (excluding differences of velocity) the time occupied is conceived as proportionate to the distance moved over. Now, the primary significance of before-and-afterness is the local one of 'in front of' and 'behind.' There it is applied to order of position; but since there is a before-and-after in magnitude, there must also be a before-and-after in movement in analogy with them. But there is also a before-and-after in time, in virtue of the dependence of time upon motion. Motion, then, is the objective seat of before-and-afterness both in movement and in time; but conceptually the before-and-afterness is distinguishable from movement.[34] Now, when we determine a movement by defining its first and last limit, we also recognize a lapse of time; for it is when we are aware of the measuring of motion by a prior and posterior limit that we may say time has passed. And our determination consists in distinguishing between the initial limit and the final one, and seeing that what lies between them is distinct from both; for when we distinguish between the extremes and what is between them, and the mind pronounces the 'nows' to be two—an initial and a final one—it is then that we say that a certain time has passed; for that which is determined either way by a 'now'[35] seems to be what we mean by time. And let this be accepted and laid down.

Accordingly, when we perceive a 'now' in isolation, that is to say not as one of two, an initial and a final one in the motion, nor yet as being a final 'now' of one period and at the same time the initial 'now' of a succeeding period, then no time seems to have elapsed, for neither has there been any corresponding motion. But when we perceive a distinct before and after, then we speak of time; for this is just what time is, the calculable measure or dimension of motion with respect to before-and-afterness.

Time, then, is not movement, but that by which movement can be numerically estimated. To see this, reflect that we estimate any kind of more-and-lessness by number; so, since we estimate all more-or-lessness on some numerical scale and estimate the more-or-lessness of motion by time, time is a scale on which something (to wit movement) can be numerically estimated. But now, since 'number' has two meanings (for we speak of the 'numbers' that are counted in the thing in question, and also of the 'numbers' by which we count them and in which we calculate), we are to note that time is the countable thing that we are counting, not the numbers we count in—which two things are different.[36]
(*Physics,* IV, i, iv, x, xi)

Local Motion

DESCARTES, from *The Principles*

René Descartes (1596–1650) was widely acclaimed among seventeenth and eighteenth century scientists, although the reason for his enormous influence did not always rest upon his successes as a scientist. In fact, the laws of collision of bodies, one of which is stated below, are obviously incorrect, and it shocked Christiaan Huyghens, the great Dutch physicist (1629–1695), to think that Descartes could publish such obvious errors. Nevertheless, it was Descartes who gave familiar expression to the Law of Inertia and who expressed his belief in the conservation of motion, though, once more, his formulation of that conservation law was incorrect. The influence he exerted upon the development of modern science lay in large measure in his systematic and varied defense of the view that the behavior of bodies and the explanation of their change of motion was to be understood by reference to principles which referred only to the shape, size, and motion of other bodies, and to nothing else. Scientists felt that at long last they knew what was the proper way to describe phenomena, and even if Descartes had been wrong in some of his principles, at least he pointed the way by indicating the kind of principles and explanations scientists ought to seek.

Principle XXV. What movement properly speaking is.

But if, looking not to popular usage, but to the truth of the matter, let us consider what ought to be understood by motion according to the truth of the thing; we may say, in order to attribute a determinate nature to it, that it is the *transference of one part of matter or one body from the vicinity of those bodies that are in immediate contact with it, and which we regard as in*

Reprinted by permission of the publisher from *Descartes' Philosophical Works*. Tr. by Haldane and Ross. Cambridge University Press, 1911, pp. 266–267.

repose, into the vicinity of others. By *one body* or by a *part of matter* I understand all that which is transported together, although it may be composed of many parts which in themselves have other motions. And I say that it is the *transportation* and not either the force or the action which transports, in order to show that the motion is always in the mobile thing, not in that which moves; for these two do not seem to me to be accurately enough distinguished. Further, I understand that it is a mode of the mobile thing and not a substance, just as figure is a mode of the figured thing, and repose of that which is at rest.

.   .   .   .   .

Principle XXXVI.

That God is the first cause of movement and that He always preserves an equal amount of movement in the universe.

Principle XXXVII.

The First law of nature: that each thing as far as in it lies, continues always in the same state; that this state changes only by *collision with other things.*

Principle XXXIX.

The Second law of nature: that all motion is of itself in a straight line; and thus things which move in a circle always tend to recede from the center of the circle which they describe.

Principle XL.

The Third law: that a body that comes into contact with another stronger than itself, loses nothing of its movement; if it meets one less strong, it loses as much as it passes over to that body.

Circular Inertia

GALILEO, from *Letters on Sunspots*

Galileo Galilei (1564–1642) discusses motion in a way which, at first sight, seems to echo Aristotle. We read in his works, for example, that some motions of bodies are natural and others are forced. He also advanced the view that only uniform circular motions are natural. These statements are surprising if they are Galileo's, for we think of him as a man who challenged the Aristotelian organization of the sciences as well as their results. Here he seems to accept both. We think of him as the discoverer of the Law of Inertia —that bodies not under the influence of forces will stay at rest (if at rest) or will move with constant speed on a straight line. Yet we now read that he believed that rectilinear motion was unnatural. However, the resemblance of his views to Aristotelian theory is misleading. Galileo used the vocabulary of natural and forced motions to express his own revolutionary discoveries. He asserted that bodies move naturally because of some intrinsic property which they possess—not because they are caused to move by some external agent or mover. Thus, Galileo retained the old distinction, but he abandoned the Aristotelian thesis, which was usually associated with it, that every motion requires an external agent.

There is no way, however, of escaping the fact that Galileo believed in a law of circular inertia, and regarded as implausible the Law of Linear Inertia which is usually ascribed to him. Nevertheless his understanding of inertia was revolutionary. Prior to his work, it was believed that the task of science was to explain every change of motion. Galileo argued that certain motions are natural, they are not caused by external agents, and that there is then no need to seek any such agents to explain their occurrence. When it was realized that motions with constant velocity are natural motions, the revolutionary consequence followed that science did not have to explain every change but only those which involved acceleration. It was Galileo's work which was responsible for the critical change in the conception of one of the tasks of physics. Galileo claimed that the Book of Nature was written in the language of mathematics and his discoveries about freely falling bodies and motion on inclined planes, some of which are discussed below, saved this insight from becoming mere rhetoric. From example after superb example of mathematical representations of physical phenomena, he drew, forcefully and elegantly, true conclusions about the world.

. . . For I seem to have observed that physical bodies have physical inclination to some motion (as heavy bodies downward), which motion is exercised by them through an intrinsic property and without need of a particular external mover, whenever they are not impeded by some obstacle. And to any other motion they have a repugnance (as the same heavy bodies to motion upward), and therefore they never move in that manner unless thrown violently by an external mover. Finally, to some movements they are indifferent, as are these same heavy bodies to horizontal motion, to which they have neither inclination(since it is not toward the center of the earth) nor repugnance (since it does not carry them away from that center). And therefore, all external impediments removed, a heavy body on a spherical surface concentric with the earth will be indifferent to rest and to movements toward any part of the horizon. And it will maintain itself in that state in which it has once been placed; that is, if placed in a state of rest, it will conserve that; and if placed in movement toward the west (for example), it will maintain itself in that movement.[1] Thus a ship, for instance, having once received some impetus through the tranquil sea, would move continually around our globe without ever stopping; and placed at rest it would perpetually remain at rest, if in the first case all extrinsic impediments could be removed, and in the second case no external cause of motion were added.

Now if this is true (as indeed it is), what would a body of ambiguous nature do if continually surrounded by an ambient that moved with a motion to which it was indifferent? I do not see how one can doubt that it would move with the motion of the ambient. And the sun, a body of spherical shape suspended and balanced upon its own center, cannot fail to follow the motion of its ambient, having no intrinsic repugnance or extrinsic impediment to rotation. It cannot have an internal repugnance, because by such a rotation it is neither removed from its place, nor are its parts permuted among themselves. Their natural arrangement is not changed in any way, so that as far as the constitution of its parts is concerned, such movement is as if it did not exist. As to external impediments, it does not seem that any obstacle can impede without contact, except perhaps by magnetic power; and in this case all that is in contact with the sun is its ambient, which not only does not impede the movement which we seek to attribute to it, but itself has this movement. This may be further confirmed, as it does not appear that any movable body can have a repugnance to a movement without having a natural propensity to the opposite motion, for in indifference no repugnance exists; hence anyone who wants to give the sun a resistance to the circular motion of its ambient would be putting in it a natural propensity for circular motion opposite to that. But this cannot appeal to any balanced mind.

Motion on Inclined Planes

GALILEO, from *The Dialogue on the Great World Systems*

. . . . .

Salv.. This principle then established, one may immediately
conclude that, if the integral components should be by their
nature movable, it is impossible that their motions should be
straight or other than circular. This reason is sufficiently easy
and manifest; for whatever moves with a straight motion changes
place and, continuing to move, moves by degrees farther away
from the term from whence it departed and from all the places
through which it has successively passed. If such motion naturally
suited with it, then it was not, in the beginning, in its proper
place; and so the parts of the world were not disposed with perfect order. But we suppose them to be perfectly ordered; therefore, as such, it is impossible that by nature they should change
place and consequently move in a straight motion. Moroeover,
the straight motion being by nature infinite, because the straight
line is infinite and indeterminate, it is impossible that any movable body can have a natural principle of moving in a straight
line, namely toward the place whither it is impossible to arrive,
there being no predetermined limit; and Nature, as Aristotle
himself well says, never attempts to do that which cannot be
done or to move whither it is impossible to arrive. And if anyone
should yet object that while the straight line, and consequently
the motion in it, can be infinitely prolonged, that is to say, is
interminate; and that yet, nevertheless, Nature has, so to speak,
arbitrarily assigned some limits, or places, and given natural
instincts to its natural bodies to move unto the same; I will reply

Reprinted from *The Dialogue on the Great World Systems* by Galileo, revised,
annotated, with an Introduction by G. de Santillana, by permission of The
University of Chicago Press, First Copyright 1953.

that this might perhaps be fabled to have come to pass in the first Chaos, where indiscrete matters confusedly and inordinately wandered. Then, to regulate them, Nature very appositely may have made use of the straight motions, by which, in the same way as the well-constituted parts, by moving, disorder themselves, so these, which were disposed chaotically, were ranged in order by this motion. But, after their exquisite distribution and collocation, it is impossible that there should remain natural inclinations in them of moving any longer in a straight motion, from which now would ensue their removal from their proper and natural place; that is to say, their disorder. We may therefore say that the straight motion serves to bring the matter into place so as to erect the work; but, once erected, it has to rest immovable or, if movable, to move only circularly.

·   ·   ·   ·   ·

Salv.: You may say, not in a year, or in ten, or in a thousand, as I will endeavor to shew you. Therefore tell me if you make any question of granting that that ball in descending goes increasing its impetus and velocity.

Sagr.: I am most certain it does.

Salv.: And if I were to say that the impetus acquired in any place of its motion is so much that it would suffice to recarry it to that place from which it came, would you grant it?

Sagr.: I should consent to it without contradiction, provided always that it might employ without impediments its whole impetus in that sole work of reconducting itself to that selfsame height, as would happen in case the Earth were bored through the centre, and the ball fell a thousand yards to the said centre, for I verily believe it would pass beyond the centre, ascending as much as it had descended. This I see plainly in the experiment of the pendulum, which, removed from the perpendicular, which is the state of rest, and afterwards let go, falls towards the said perpendicular, and goes as far beyond it, or only so much less as the opposition of the air, and line, or other such accidents have hindered it. The like I see in the water, which, descending through a pipe, ascends as much as it has descended.

Salv.: You argue very well. And as I know that you will not scruple to grant that the gain in the impetus goes with the recession from the point where the body departed, and its approach to the centre whither its motions tend, would you grant that two equal bodies though descending by different lines, without any impediment, acquire equal impetus, provided that the distances to the centre be equal?

Sagr.: I do not very well understand the question.

Salv.: I will express it better by drawing a figure. I will suppose the line AB (in Fig. 1) parallel to the horizon, and upon point B I will erect a perpendicular BC; and after that I add this slanting line CA. CA is an inclined plane exquisitely polished and hard, upon which descends a ball perfectly round and of a very hard substance. Such another ball I suppose to descend freely along the perpendicular CB. Will you now admit that the impetus of that which decends along the plane CA, being arrived at A, may be equal to the impetus acquired by the other in the point B, after the descent along the perpendicular CB?

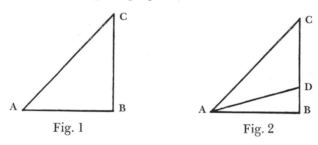

Fig. 1   Fig. 2

Sagr.: I resolutely believe so; for in effect they both have the same proximity to the centre, and, by what I have already granted, their impetuosities would be equally sufficient to carry them back to the same height.

. . . . .

Going back to the same figure we used before [Fig. 1] let us remember that we agreed that the body descending along CA and the falling one along CB were found to have acquired equal degrees of velocity at the points B and A. Now to proceed, I suppose you will not object to grant that upon another plane less steep than AC, as, for example, AD (in Fig. 2), the motion of the descending body would be yet more slow than on the plane AC. So that it is not the least doubtful that there may be planes so little elevated above the horizontal AB that the same ball will take as long a time as we choose to reach A, whereas, to reach that horizontally, an infinite time will not suffice: the less steep AD, the longer is the time necessary for the descent. It must be said that there is a point taken on CB so near to B that, drawing a plane through it to the point A, the ball would not traverse it in a whole year. You should know next that, if the ball continued to travel from the point A, without accelerating or retarding, for as long a time again as it was in coming down the inclined plane, it would pass double the length of that plane. For example, if the ball had passed the plane DA in an hour, then, continuing to move uniformly with that velocity it is found to have acquired by the term A, it will pass in another hour a space double the length of DA; and because (as we have said) the velocities acquired at the points B and A by the bodies that depart from any point taken along CB, and that descend one by DC, the other by CB, are always equal; therefore, the body falling by the perpendicular may depart from a point so near to B that the velocity, held uniform, acquired at B would not suffice to conduct the body a distance double the length of the inclined plane in a year, or in ten, or a hundred.

In brief, all external and accidental impediments removed, a body naturally moves upon an inclined plane with greater and greater slowness, according as the inclination is less, so that in the end the slowness comes to be infinite when the inclined plane coincides with the horizontal; also, the velocity acquired at some point on the inclined plane is equal to the velocity of the body that descends by CB to the point cut by a parallel to the horizon which passes through that point of the inclined plane. Hence, it must be granted that the body in falling any distance in a straight line passes through all the infinite degrees of slowness, acquiring a definite velocity determined by the inclination of the plane. There can be a plane with such a small inclination that the body must traverse a great distance over a long time before acquiring the assigned velocity; and along the horizontal plane the body would, of course, never gain any velocity, as it will never move by itself; but this motion by the horizontal line, which is neither declined nor inclined, is a circular motion. . . . Therefore, a circular motion is never acquired naturally, without being preceded by rectilinear motion; but, once acquired, it will continue perpetually with uniform velocity. I could demonstrate the same truth by other means, but I will not by so great a digression interrupt our principal argument; rather, I will return to the subject on some other occasion. . . .

Motion: Uniform and Accelerated

GALILEO, from *The Dialogue Concerning Two New Sciences*

. . . . .

Change of Position. [De Motu Locali]

My purpose is to set forth a very new science dealing with a very ancient subject. There is, in nature, perhaps nothing older than motion, concerning which the books written by philosophers are neither few nor small; nevertheless I have discovered by experiment some properties of it which are worth knowing and which have not hitherto been either observed or demonstrated. Some superficial observations have been made, as, for instance, that the free motion[1] [*naturalem motum*] of a heavy falling body is continuously accelerated; but to just what extent this acceleration occurs has not yet been announced; for so far as I know, no one has yet pointed out that the distances traversed, during equal intervals of time, by a body falling from rest, stand to one another in the same ratio as the odd numbers beginning with unity.[2]

It has been observed that missiles and projectiles describe a curved path of some sort; however no one has pointed out the fact that this path is a parabola. But this and other facts, not few in number or less worth knowing, I have succeeded in proving; and what I consider more important, there have been opened up to this vast and most excellent science, of which my work is merely the beginning, ways and means by which other minds more acute than mine will explore its remote corners.

This discussion is divided into three parts; the first part deals

Reprinted by permission of Northwestern University Press, Evanston, Illinois, from *The Dialogue Concerning Two New Sciences*, tr. by H. Crew and A. de Salvio. Originally published by The Macmillan Company, 1914.

51

with motion which is steady or uniform; the second treats of motion as we find it accelerated in nature; the third deals with the so-called violent motions and with projectiles.

Uniform Motion

In dealing with steady or uniform motion, we need a single definition which I give as follows:

DEFINITION

By steady or uniform motion, I mean one in which the distances traversed by the moving particle during any equal intervals of time, are themselves equal.

CAUTION

We must add to the old definition (which defined steady motion simply as one in which equal distances are traversed in equal times) the word "any," meaning by this, all equal intervals of time; for it may happen that the moving body will traverse equal distances during some equal intervals of time and yet the distances traversed during some small portion of these time-intervals may not be equal, even though the time-intervals be equal.

From the above definition, four axioms follow, namely:

Axiom I

In the case of one and the same uniform motion, the distance traversed during a longer interval of time is greater than the distance traversed during a shorter interval of time.

Axiom II

In the case of one and the same uniform motion, the time required to traverse a greater distance is longer than the time required for a less distance.

Axiom III

In one and the same interval of time, the distance traversed at a greater speed is larger than the distance traversed at a less speed.

Axiom IV

The speed required to traverse a longer distance is greater than that required to traverse a shorter distance during the same time-interval.

Theorem I, Proposition I

If a moving particle, carried uniformly at a constant speed, traverses two distances the time-intervals required are to each other in the ratio of these distances.

Naturally Accelerated Motion

The properties belonging to uniform motion have been discussed in the preceding section; but accelerated motion remains to be considered.

And first of all it seems desirable to find and explain a definition best fitting natural phenomena. For anyone may invent an arbitrary type of motion and discuss its properties; thus, for instance, some have imagined helices and conchoids as described by certain motions which are not met in nature, and have very commendably established the properties which these curves possess in virtue of their definitions; but we have decided to consider the phenomena of bodies falling with an acceleration such as actually occurs in nature and to make this definition of accelerated motion exhibit the essential features of observed accelerated motions. And this, at last, after repeated efforts we trust we have succeeded in doing. In this belief we are confirmed mainly by the consideration that experimental results are seen to agree with and exactly correspond with those properties which have been, one after another, demonstrated by us. Finally, in the investigation of naturally accelerated motion we were led, by hand as it were, in following the habit and custom of nature herself, in all her various other processes, to employ only those means which are most common, simple and easy.

For I think no one believes that swimming or flying can be accomplished in a manner simpler or easier than that instinctively employed by fishes and birds.

When, therefore, I observe a stone initially at rest falling from an elevated position and continually acquiring new increments of speed, why should I not believe that such increases take place in a manner which is exceedingly simple and rather obvious to everybody? If now we examine the matter carefully we find no addition or increment more simple than that which repeats itself always in the same manner. This we readily understand when we consider the intimate relationship between time and motion; for just as uniformity of motion is defined by and conceived through equal times and equal spaces (thus we call a motion uniform when equal distances are traversed during equal time-intervals), so also we may, in a similar manner, through equal time-intervals, conceive additions of speed as taking place without complication; thus we may picture to our mind a motion as uniformly and continuously accelerated when, during any equal intervals of time whatever, equal increments of speed are given to it. Thus if any equal intervals of time whatever have elapsed, counting from that time at which the moving body left its position of rest and began to descend, the amount of speed acquired during the first two time-intervals will be double that acquired during the first time-interval alone; so the amount added during three of these time-intervals will be treble; and that in four, quadruple that of the first time-interval. To put the matter more clearly, if a body were to continue its motion with the same speed which it had acquired during the first time-interval and were to retain this same uniform speed, then its motion would be twice as slow as that which it would have if its velocity had been acquired during *two* time-intervals.

And thus, it seems, we shall not be far wrong if we put the increment of speed as proportional to the increment of time; hence the definition of motion which we are about to discuss may be stated as follows: A motion is said to be uniformly accelerated, when starting from rest, it acquires, during equal time-intervals, equal increments of speed.

.   .   .   .   .

Salv.: The present does not seem to be the proper time to investigate the cause of the acceleration of natural motion concerning which various opinions have been expressed by various philosophers, some explaining it by attraction to the center, others to repulsion between the very small parts of the body, while still others attribute it to a certain stress in the surrounding medium which closes in behind the falling body and drives it from one of its positions to another. Now, all these fantasies, and others too, ought to be examined; but it is not really worth while. At present it is the purpose of our Author merely to investigate and to demonstrate some of the properties of accelerated motion (whatever the cause of this acceleration may be)—meaning thereby a motion, such that the momentum of its velocity [*i momenti della sua velocita*] goes on increasing after departure from rest, in simple proportionality to the time, which is the same as saying that in equal time-intervals the body receives equal increments of velocity; and if we find the properties [of accelerated motion] which will be demonstrated later are realized in freely falling and accelerated bodies, we may conclude that the assumed definition includes such a motion of falling bodies and that their speed [*accelerazione*] goes on increasing as the time and the duration of the motion.

.   .   .   .   .

Sagr.: But now, continuing the thread of our talk, it would seem that up to the present we have established the definition of uniformly accelerated motion which is expressed as follows:

A motion is said to be equally or uniformly accelerated when, starting from rest, its momentum (*celeritatis momenta*) receives equal increments in equal times.

Salv.: This definition established, the Author makes a single assumption, namely,

The speeds acquired by one and the same body moving down planes of different inclinations are equal when the heights of these planes are equal.

By the height of an inclined plane we mean the perpendicular let fall from the upper end of the plane upon the horizontal line drawn through the lower end of the same plane. Thus, to illustrate [Fig. 1], let the line AB be horizontal, and let the planes CA and CD be inclined to it; then the Author calls the perpendicular CB the "height" of the planes CA and CD; he supposes that the speeds acquired by one and the same body, descending along the planes CA and CD to the terminal points A and D are equal since the heights of these planes are the same, CB; and also it must be understood that this speed is that which would be acquired by the same body falling from C to B.

Sagr.: Your assumption appears to me so reasonable that it ought to be conceded without question, provided of course there

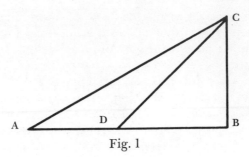

Fig. 1

are no chance or outside resistances, and that the planes are hard
and smooth, and that the figure of the moving body is perfectly
round, so that neither plane nor moving body is rough. All
resistance and opposition having been removed, my reason tells
me at once that a heavy and perfectly round ball descending
along the lines CA, CD, CB would reach the terminal points A,
D, B, with equal momenta [*impeti eguali*].

Salv.: Your words are very plausible; but I hope by experience
to increase the probability to an extent which shall be little
short of a rigid demonstration.

Imagine this page to represent a vertical wall, with a nail
driven into it; and from the nail let there be suspended a lead
bullet [Fig. 2] of one or two ounces by means of a fine vertical
thread, AB, say from four to six feet long; on this wall draw a
horizontal line DC, at right angles to the vertical thread AB,
which hangs about two finger-breadths in front of the wall. Now

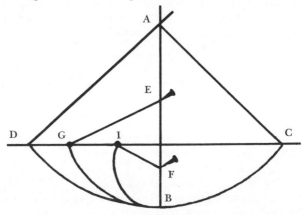

Fig. 2

bring the thread AB with the attached ball into the position AC
and set it free; first it will be observed to descend along the arc
CBD, to pass the point B, and to travel along the arc BD, till it
almost reaches the horizontal CD, a slight shortage being caused
by the resistance of the air and the string; from this we may
rightly infer that the ball in its descent through the arc CB acquired a momentum [*impeto*] on reaching B, which was just
sufficient to carry it through a similar arc BD to the same height.
Having repeated this experiment many times, let us now drive
a nail into the wall close to the perpendicular AB, say at E or F, so
that it projects out some five or six finger-breadths in order that the
thread, again carrying the bullet through the arc CB, may strike
upon the nail E when the bullet reaches B, and thus compel
it to traverse the arc BG, described about E as center. From this
we can see what can be done by the same momentum [*impeto*]
which previously starting at the same point B carried the same
body through the arc BD to the horizontal CD. Now, gentlemen,
you will observe with pleasure that the ball swings to the point
G in the horizontal, and you would see the same thing happen if
the obstacle were placed at some lower point, say at F, about
which the ball would describe the arc BI, the rise of the ball
always terminating exactly on the line CD. But when the nail is
placed so low that the remainder of the thread below it will not
reach to the height CD (which would happen if the nail were
placed nearer B than to the intersection of AB with the horizontal CD) then the thread leaps over the nail and twists itself
about it.

This experiment leaves no room for doubt as to the truth of
our supposition; for since the two arcs CB and DB are equal and
similarly placed, the momentum [*momento*] acquired by the fall
through the arc CB is the same as that gained by fall through the
arc DB; but the momentum [*momento*] acquired at B, owing to
fall through CB, is able to lift the same body [*mobile*] through
the arc BD; therefore, the momentum acquired in the fall BD
is equal to that which lifts the same body through the same arc
from B to D; so, in general, every momentum acquired by fall
through the arc is equal to that which can lift the same body
through the same arc. But all these momenta [*momenti*] which
cause a rise through the arcs BD, BG, and BI are equal, since
they are produced by the same momentum, gained by fall through

CB, as experiment shows. Therefore all the momenta gained by fall through the arcs DB, GB, IB are equal.

Sagr.: The argument seems to me so conclusive and the experiment so well adapted to establish the hypothesis that we may, indeed, consider it as demonstrated.

Salv.: I do not wish, Sagredo, that we trouble ourselves too much about this matter, since we are going to apply this principle mainly in motions which occur on plane surfaces, and not upon curved, along which acceleration varies in a manner greatly different from that which we have assumed for planes.

So that, although the above experiment shows us that the descent of the moving body through the arc CB confers upon it momentum [*momento*] just sufficient to carry it to the same height through any of the arcs BD, BG, BI, we are not able, by similar means, to show that the event would be identical in the case of a perfectly round ball descending along planes whose inclinations are respectively the same as the chords of these arcs. It seems likely, on the other hand, that, since these planes form angles at the point B, they will present an obstacle to the ball which has descended along the chord CB, and starts to rise along the chord BD, BG, BI.

In striking these planes some of its momentum [*impeto*] will be lost and it will not be able to rise to the height of the line CD; but this obstacle, which interferes with the experiment, once removed, it is clear that the momentum [*impeto*] (which gains in strength with descent) will be able to carry the body to the same height. Let us then, for the present, take this as a postulate, the absolute truth of which will be established when we find that the inferences from it correspond to and agree perfectly with experiment. The Author having assumed this single principle passes next to the propositions which he clearly demonstrates; the first of these is as follows:

Theorem I, Proposition I

The time in which any space is traversed by a body starting from rest and uniformly accelerated is equal to the time in which that same space would be traversed by the same body moving at a uniform speed whose value is the mean of the highest speed and the speed just before acceleration began.

Let us represent by the line AB [Fig. 3] the time in which the space CD is traversed by a body which starts from rest at C and is uniformly accelerated; let the final and highest value of the speed gained during the interval AB be represented by the line EB drawn at right angles to AB; draw the line AE; then all lines drawn from equidistant points on AB and parallel to BE will

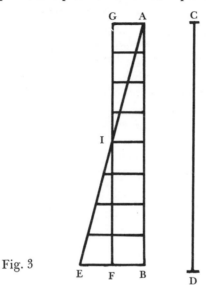

Fig. 3

represent the increasing values of the speed, beginning with the instant A. Let the point F bisect the line EB; draw FG parallel to BA, and GA parallel to FB, thus forming a parallelogram AGFB which will be equal in area to the triangle AEB, since the side GF bisects the side AE at the point I; for if the parallel lines in the triangle AEB are extended to GI, then the sum of all the parallels contained in the quadrilateral is equal to the sum of those contained in the triangle AEB; for those in the triangle IEF are equal to those contained in the triangle GIA, while those included in the trapezium AIFB are common. Since each and every instant of time in the time-interval AB has its corresponding point on the line AB, from which points parallels drawn in and limited by the traingle AEB represent the increasing values of the growing velocity, and since parallels contained within the rectangle represent the values of a speed which is not increasing, but constant, it appears, in like manner, that the momenta

[*momenta*] assumed by the moving body may also be represented, in the case of the accelerated motion, by the increasing parallels of the triangle AEB, and, in the case of the uniform motion, by the parallels of the rectangle GB. For, what the momenta may lack in the first part of the accelerated motion (the deficiency of the momenta being represented by the parallels of the triangle AGI) is made up by the momenta represented by the parallels of the triangle IEF.

Hence it is clear that equal spaces will be traversed in equal times by two bodies, one of which, starting from rest, moves with a uniform acceleration, while the momentum of the other, moving with uniform speed, is one-half its maximum momentum under accelerated motion.

Q.E.D.

.   .   .   .   .

Sagr.: Please suspend the discussion for a moment since there just occurs to me an idea which I want to illustrate by means of a diagram in order that it may be clearer both to you and to me.

Let the line AI [Fig. 4] represent the lapse of time measured from the initial instant A; through A draw the straight line AF

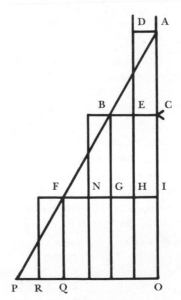

Fig. 4

making any angle whatever; join the terminal points I and F; divide the time AI in half at C; draw CB parallel to IF. Let us consider CB as the maximum value of the velocity which increases from zero at the beginning, in simple proportionality to the intercepts on the triangle ABC of lines drawn parallel to BC; or what is the same thing, let us suppose the velocity to increase in proportion to the time; then I admit without question, in view of the preceding argument, that the space described by a body falling in the aforesaid manner will be equal to the space traversed by the same body during the same length of time travelling with a uniform speed equal to EC, the half of BC. Further let us imagine that the body has fallen with accelerated motion so that, at the instant C, it has the velocity BC. It is clear that if the body continued to descend with the same speed BC, without acceleration, it would in the next time-interval CI traverse double the distance covered during the interval AC, with the uniform speed EC which is half of BC; but since the falling body acquires equal increments of speed during equal increments of time, it follows that the velocity BC, during the next time-interval CI will be increased by an amount represented by the parallels of the triangle BFG which is equal to the triangle ABC. If, then, one adds to the velocity GI half of the velocity FG, the highest speed acquired by the accelerated motion and determined by the parallels of the triangle BFG, he will have the uniform velocity with which the same space would have been described in the time CI; and since this speed IN is three times as great as EC it follows that the space described during the interval CI is three times as great as that described during the interval AC. Let us imagine the motion extended over another equal time-interval IO, and the triangle extended to APO; it is then evident that if the motion continues during the interval IO, at the constant rate IF acquired by acceleration during the time AI, the space traversed during the interval IO will be four times that traversed during the first interval AC, because the speed IF is four times the speed EC. But if we enlarge our triangle so as to include FPQ which is equal to ABC, still assuming the acceleration to be constant, we shall add to the uniform speed an increment RQ, equal to EC; then the value of the equivalent uniform speed during the time-interval IO will be five times that during the first time-interval AC; therefore the space traversed will be quintuple

that during the first interval AC. It is thus evident by simple
computation that a moving body starting from rest and acquiring
velocity at a rate proportional to the time, will, during equal
intervals of time, traverse distances which are related to each
other as the odd numbers beginning with unity, 1, 3, 5;[3] or
considering the total space traversed, that covered in double time
will be quadruple that covered during unit time; in triple time,
the space is nine times as great as in unit time. And in general the
spaces traversed are in the duplicate ratio of the time, i.e., in the
ratio of the squares of the times.

Simp.: In truth, I find more pleasure in this simple and clear
argument of Sagredo than in the Author's demonstration which
to me appears rather obscure; so that I am convinced that matters
are as described, once having accepted the definition of uniformly
accelerated motion. But as to whether this acceleration is that
which one meets in nature in the case of falling bodies, I am
still doubtful; and it seems to me, not only for my own sake but
also for all those who think as I do, that this would be the proper
moment to introduce one of those experiments—and there are
many of them, I understand—which illustrate in several ways the
conclusions reached.

Salv.: The request which you, as a man of science, make, is a
very reasonable one; for this is the custom—and properly so—
in those sciences where mathematical demonstrations are applied
to natural phenomena, as is seen in the case of perspective, astronomy, mechanics, music, and others where the principles, once
established by well-chosen experiments, become the foundations
of the entire superstructure. I hope therefore it will not appear to
be a waste of time if we discuss at considerable length this first
and most fundamental question upon which hinge numerous
consequences of which we have in this book only a small number,
placed there by the Author, who has done so much to open a
pathway hitherto closed to minds of speculative turn. So far as
experiments go they have not been neglected by the Author; and
often, in his company, I have attempted in the following manner
to assure myself that the acceleration actually experienced by
falling bodies is that above described.

A piece of wooden moulding or scantling, about 12 cubits long,
half a cubit wide, and three finger-breadths thick, was taken;
on its edge was cut a channel a little more than one finger in

breadth; having made this groove very straight, smooth, and polished, and having lined it with parchment, also as smooth and polished as possible, we rolled along it a hard, smooth, and very round bronze ball. Having placed this board in a sloping position, by lifting one end some one or two cubits above the other, we rolled the ball, as I was just saying, along the channel, noting, in a manner presently to be described, the time required to make the descent. We repeated this experiment more than once in order to measure the time with an accuracy such that the deviation between two observations never exceeded one-tenth of a pulse-beat. Having performed this operation and having assured ourselves of its reliability, we now rolled the ball only one-quarter the length of the channel; and having measured the time of its descent, we found it precisely one-half of the former. Next we tried other distances, comparing the time for the whole length with that for the half, or with that for two-thirds, or three-fourths, or indeed for any fraction; in such experiments, repeated a full hundred times, we always found that the spaces traversed were to each other as the squares of the times, and this was true for all inclinations of the plane, i.e., of the channel, along which we rolled the ball. We also observed that the times of descent, for various inclinations of the plane, bore to one another precisely that ratio which, as we shall see later, the Author had predicted and demonstrated for them.

For the measurement of time, we employed a large vessel of water placed in an elevated position; to the bottom of this vessel was soldered a pipe of small diameter giving a thin jet of water, which we collected in a small glass during the time of each descent, whether for the whole length of the channel or for a part of its length; the water thus collected was weighed, after each descent, on a very accurate balance; the differences and ratios of these weights gave us the differences and ratios of the times, and this with such accuracy that although the operation was repeated many, many times, there was no appreciable discrepancy in the results.

Simp.: I would like to have been present at these experiments; but feeling confidence in the care with which you performed them, and in the fidelity with which you relate them, I am satisfied and accept them as true and valid.

Absolute Space, Time, and the Science of Motion

ISAAC NEWTON, from *The Mathematical Principles of Natural Philosophy and the System of the World*

Every schoolboy has some acquaintance with the amazing work of Sir Isaac Newton (1642–1727). With the aid of his three laws of motion, together with the Law of Universal Gravitation, he managed to explain the laws of planetary motion, the intricate motion of the moon, tides, the velocity of sound in air, Galileo's Law of Falling Bodies, and a host of less well-known phenomena. His theory of motion was the first example of a theory which explained and systematized various independently known laws and effects, showing why they are true, or within what margin they are sound.

For many of Newton's contemporaries, however, his theory was a mixed blessing. So mixed in fact that there was still strong opposition to its acceptance even fifty years after its publication (1687). More than two hundred years later, the philosopher-physicist Ernst Mach (see p. 139) called the theory physically meaningless and suggested its modification. The truth seems to be that up to the middle of the nineteenth century, the major objections to Newton's formulation of mechanics did not question its accuracy or its scope. For various reasons, the use of concepts like mutual attraction, matter, Absolute Space and Absolute Time, which Newton referred to in his formulation of the science of mechanics, came under attack.

Absolute Space and Time are the hallmark of Newton's theory of motion. He believed that there is a unique entity, which he called Absolute Space, which has parts to it. (We shall not discuss Absolute Time, which is also an entity according to Newton, whose parts, unlike those of Absolute Space, are not simultaneous.) This entity is not a superbody, but bodies moved in it. He agreed that bodies move with respect to each other—who except Zeno

Reprinted by permission of the publisher from *The Mathematical Principles of Natural Philosophy and the System of the World* by Isaac Newton, tr. by A. Motte, F. Cajori, ed. (Berkeley: University of California Press, 1947).

would deny this! But the true laws of motion do not refer to the relative motion of bodies. True motion occurs when a body changes its place in Absolute Space. And the true laws, for example, his three laws of motion, concern changes which bodies undergo when they move with respect to Absolute Space. Newton also believed that certain kinds of motion (rotations, for example) produce certain forces which act on the moving body and distort its shape or effect some other change in it or its motion. Thus it is apparent that Absolute Space is, according to Newton, not merely an entity which serves to describe the true positions and motions of bodies; it has a causal efficacy as well.

Newton was very well aware of the fact that both Absolute Space and Time are not observable entities. It is impossible, as he remarked, to see whether a body is moving with respect to Absolute Space, in the way in which one can see two bodies moving with respect to each other. Nevertheless he thought that it is sometimes possible to show that a body is truly rotating with respect to Absolute Space. He offered two celebrated experiments, the rotating bucket experiment and the case of two rotating bodies connected by a taut string. (For a critical discussion of the former, see Mach, p. 139.) It is true that neither of these experiments showed conclusively that a body was rotating with respect to Absolute Space. But it is important to remember that Newton believed that the existence of motion with respect to Absolute Space is an empirical claim, which is supportable by empirical evidence. Newton did not believe in a narrow empiricism, according to which genuine claims about the world could refer only to those entities and properties which are accessible to sensory experience. He was willing, as we have seen, to argue for certain claims which involve entities like Absolute Space, which are simply not to be seen, touched, smelled, or heard.

The reader may be struck by Newton's description of certain rules of reasoning in philosophy. "Why," it might be asked, "once he was in possession of his theory of motion, did he require the specification of rules for doing science? Were his contemporaries so lacking in understanding, that a mere statement of the theory was insufficient?"

The answer has to be gleaned from the remarks of Newton's friends. Newton realized that, contrary to what Descartes had said, the science of mechanics is different from geometry both in the kind of concepts referred to and the kind of support provided for its propositions. Newton therefore had to delineate the kinds of concepts which belong to the science of mechanics. Since the laws of motion could not be given the same kind of "proof" as the truths of geometry, Newton also had to say something about the rules of acceptance (or rejection) of propositions of science. His rules of reasoning were designed to answer both kinds of problems.

RULES OF REASONING IN PHILOSOPHY

Rule I

We are to admit no more causes of natural things than such as are both true and sufficient to explain their appearances.

To this purpose the philosophers say that Nature does nothing in vain, and more is in vain when less will serve; for Nature is pleased with simplicity, and affects not the pomp of superfluous causes.

Rule II

Therefore to the same natural effects we must, as far as possible, assign the same causes.

As to respiration in a man and in a beast; the descent of stones in *Europe* and in *America;* the light of our culinary fire and of the sun; the reflection of light in the earth, and in the planets.

Rule III

The qualities of bodies, which admit neither intensification nor remission of degrees, and which are found to belong to all bodies within the reach of our experiments, are to be esteemed the universal qualities of all bodies whatsoever.

For since the qualities of bodies are only known to us by experiments, we are to hold for universal all such as universally agree with experiments; and such as are not liable to diminution can never be quite taken away. We are certainly not to relinquish the evidence of experiments for the sake of dreams and vain fictions of our own devising; nor are we to recede from the analogy of Nature, which is wont to be simple, and always consonant to itself. We no other way know the extension of bodies than by our senses, nor do these reach it in all bodies; but because we perceive extension in all that are sensible, therefore we ascribe it universally to all others also. That abundance of bodies are hard, we learn by experience; and because the hardness of the whole arises from the hardness of the parts, we therefore justly infer the hardness of the undivided particles not only of the bodies we feel but of all others. That all bodies are impenetrable, we gather not from reason, but from sensation. The bodies which we handle we find impenetrable, and thence conclude impenetrability to be an universal property of all bodies whatsoever. That all bodies are movable, and endowed with certain powers (which we call the inertia) of persevering in their motion, or in their rest, we only infer from the like properties observed in the bodies which we have seen. The extension, hardness, impenetrability, mobility, and inertia of the whole, result from the extension, hardness, impenetrability, mobility, and inertia of the parts; and hence we conclude the least particles of all bodies to be also all extended, and hard and impenetrable, and movable, and endowed with their proper inertia. And this is the foundation of all philosophy. Moreover, that the divided but contiguous particles of bodies may be separated from one another, is matter of observation; and, in the particles that remain undivided, our minds are able to distinguish yet lesser parts, as is mathematically demonstrated. But whether the parts so distinguished, and not yet divided, may, by the powers of Nature, be actually divided and separated from one another, we cannot certainly determine. Yet, had we the proof of but one experiment that any undivided particle, in breaking a hard and solid body, suffered a division, we might by virtue of this rule conclude that the undivided as well as the divided particles may be divided and actually separated to infinity.

Lastly, if it universally appears, by experiments and astronomical observations, that all bodies about the earth gravitate towards the earth, and that in proportion to the quantity of matter which they severally contain; that the moon likewise, according to the quantity of its matter, gravitates towards the earth; that, on the other hand, our sea gravitates towards the moon; and all the planets one towards another; and the comets in like manner towards the sun; we must, in consequence of this rule, universally allow that all bodies whatsoever are endowed with the principle of mutual gravitation. For the argument from the appearances concludes with more force for the universal gravitation of all bodies than for their impenetrability; of which, among those in the celestial regions, we have no experiments, nor any manner of observation. Not that I affirm gravity to be essential to bodies: by their *vis insita* I mean nothing but their inertia. This is immutable. Their gravity is diminished as they recede from the earth.

Rule IV

In experimental philosophy we are to look upon propositions inferred by general induction from phenomena as accurately or very nearly true, notwithstanding any contrary hypotheses that may be imagined, till such time as other phenomena occur, by which they may either be made more accurate or liable to exceptions.

This rule we must follow, that the argument of induction may not be evaded by hypotheses.

MATHEMATICAL PRINCIPLES OF NATURAL PHILOSOPHY

Definitions

Definition 1

The quantity of matter is the measure of the same, arising from its density and bulk conjointly.

Thus air of a double density, in a double space, is quadruple in quantity; in a triple space, sextuple in quantity. The same thing is to be understood of snow, and fine dust or powders, that are condensed by compression or liquefaction, and of all bodies that are by any causes whatever differently condensed. I have no regard in this place to a medium, if any such there is, that freely pervades the interstices between the parts of bodies. It is this quantity that I mean hereafter everywhere under the name of body or mass. And the same is known by the weight of each body, for it is proportional to the weight, as I have found by experiments on pendulums, very accurately made, which shall be shown hereafter.

Definition II

The quantity of motion is the measure of the same, arising from the velocity and quantity of matter conjointly.

The motion of the whole is the sum of the motions of all the parts; and therefore in a body double in quantity, with equal velocity, the motion is double; with twice the velocity, it is quadruple.

Definition III

The vis insita, *or innate force of matter, is a power of resisting, by which every body, as much as in it lies, continues in its present state, whether it be of rest, or of moving uniformly forwards in a right line.*

This force is always proportional to the body whose force it is and differs nothing from the inactivity of the mass, but in our manner of conceiving it. A body, from the inert nature of matter, is not without difficulty put out of its state of rest or motion. Upon which account, this *vis insita* may, by a most significant name, be called inertia (*vis inertiæ*) or force of inactivity. But a body only exerts this force when another force, impressed upon it, endeavors to change its condition; and the exercise of this force may be considered as both resistance and impulse; it is resistance so far as the body, for maintaining its present state, opposes the force impressed; it is impulse so far as the body, by not easily giving way to the impressed force of another, endeavors to change the state of that other. Resistance is usually ascribed to bodies at rest, and impulse to those in motion; but motion and rest, as commonly conceived, are only relatively distinguished; nor are those bodies always truly at rest, which commonly are taken to be so.

Definition IV

An impressed force is an action exerted upon a body, in order to change its state, either of rest, or of uniform motion in a right line.

This force consists in the action only, and remains no longer in the body when the action is over. For a body maintains every new state it acquires, by its inertia only. But impressed forces are of different origins, as from percussion, from pressure, from centripetal force.

Definition V

A centripetal force is that by which bodies are drawn or impelled, or any way tend, towards a point as to a centre.

Of this sort is gravity, by which bodies tend to the centre of the earth; magnetism, by which iron tends to the loadstone; and that force, whatever it is, by which the planets are continually drawn aside from the rectilinear motions, which otherwise they would pursue, and made to revolve in curvilinear orbits. A stone, whirled about in a sling, endeavors to recede from the hand that turns it; and by that endeavor, distends the sling, and that with so much the greater force, as it is revolved with the greater velocity, and as soon as it is let go, flies away. That force which opposes itself to this endeavor, and by which the sling continually draws back the stone towards the hand, and retains it in its orbit, because it is directed to the hand as the centre of the orbit, I call the centripetal force. And the same thing is to be understood of all bodies, revolved in any orbits. They all endeavor to recede from the centres of their orbits; and were it not for the opposition of a contrary force which restrains them to, and detains them in their orbits, which I therefore call centripetal, would fly off in right lines, with an uniform motion. A projectile, if it was not for the force of gravity, would not deviate towards the earth, but would go off from it in a right line, and that with an uniform motion, if the resistance of the air was taken away. It is by its gravity that it is drawn aside continually from it rectilinear course, and made to deviate towards the earth, more or less, according to the force of its gravity, and the velocity of its motion. The less its gravity is, or the quantity of its matter, or the greater the velocity with which it is projected, the less will it deviate from a rectilinear course, and the farther it will go. If a leaden ball, projected from the top of a mountain by the force of gunpowder, with a given velocity, and in a direction parallel to the horizon, is carried in a curved line to the distance of two miles before it falls to the ground; the same, if the resistance of the air were taken away, with a double or decuple velocity, would fly twice or ten times as far. And by increasing the velocity, we may at pleasure increase the distance to which it might be projected, and diminish the curvature of the line which it might describe, till at last it should fall at the distance of 10, 30, or 90 degrees, or even might go quite round the whole earth before it falls; or lastly, so that it might never fall to the earth, but go forwards into the celestial spaces, and proceed in its motion *in infinitum.* And after the same manner that a projectile, by the force of gravity, may be made to revolve in an orbit, and go round the whole earth, the moon also, either by the force of gravity, if it is endued with gravity, or by any other force, that impels it towards the earth, may be continually drawn aside towards the earth, out of the rectilinear way which by its innate force it would pursue; and would be made to revolve in the orbit which it now describes; nor could the moon without some such force be retained in its orbit. If this force was too small, it would not sufficiently turn the moon out of a rectilinear course; if it was too great, it would turn it too much, and draw down the moon from its orbit towards the earth. It is necessary that the force be of a just quantity, and it belongs to the mathematicians to find the force that may serve exactly to retain a body in a given orbit with a given velocity; and *vice versa,* to determine the curvilinear way into which a body projected from a given place, with a given velocity, may be made to deviate from it natural rectilinear way, by means of a given force.

The quantity of any centripetal force may be considered as of three kinds: absolute, accelerative, and motive.

Definition VI

The absolute quantity of a centripetal force is the measure of the same, proportional to the efficacy of the cause that propagates it from the centre, through the spaces round about.

Thus the magnetic force is greater in one loadstone and less in another, according to their sizes and strength of intensity.

Definition VII

The accelerative quantity of a centripetal force is the measure of the same, proportional to the velocity which it generates in a given time.

Thus the force of the same loadstone is greater at a less distance, and less at a greater: also the force of gravity is greater in valleys, less on tops of exceeding high mountains; and yet less (as shall hereafter be shown), at greater distances from the body of the earth; but at equal distances, it is the same everywhere; because (taking away, or allowing for, the resistance of the air), it equally accelerates all falling bodies, whether heavy or light, great or small.

Definition VIII

The motive quantity of a centripetal force is the measure of the same, proportional to the motion which it generates in a given time.

Thus the weight is greater in a greater body, less in a less body; and, in the same body, it is greater near to the earth, and less at remoter distances. This sort of quantity is the centripetency, or propension of the whole body towards the centre, or, as I may say, its weight; and it is always known by the quantity of an equal and contrary force just sufficient to hinder the descent of the body.

These quantities of forces, we may, for the sake of brevity, call by the names of motive, accelerative, and absolute forces; and, for the sake of distinction, consider them with respect to the bodies that tend to the centre, to the places of those bodies, and to the centre of force towards which they tend; that is to say, I refer the motive force to the body as an endeavor and propensity of the whole towards a centre, arising from the propensities of the several parts taken together; the accelerative force to the place of the body, as a certain power diffused from the centre to all places around to move the bodies that are in them; and the absolute force to the centre, as endued with some cause, without which those motives forces would not be propogated through the spaces round about; whether that cause be some central body (such as is the magnet in the centre of the magnetic force, or the earth in the centre of the gravitating force), or anything else that does not yet appear. For I here design only to give a mathematical notion of those forces, without considering their physical causes and seats.

Wherefore the accelerative force will stand in the same relation to the motive, as celerity does to motion. For the quantity of motion arises from the celerity multiplied by the quantity of matter; and the motive force arises from the accelerative force multiplied by the same quantity of matter. For the sum of the actions of the accelerative force, upon the several particles of the body, is the motive force of the whole. Hence it is, that near the surface of the earth, where the accelerative gravity, or force productive of gravity, in all bodies is the same, the motive gravity or the weight is as the body; but if we should ascend to higher regions, where the accelerative gravity is less, the weight would be equally diminished, and would always be as the product of the body, by the accelerative gravity. So in those regions, where the accelerative gravity is diminished into one-half, the weight of a body two or three times less, will be four or six times less.

I likewise call attractions and impulses, in the same sense, accelerative, and motive; and use the words attraction, impulse, or propensity of any sort towards a centre, promiscuously, and in-differently, one for another; considering those forces not phys-ically, but mathematically; wherefore the reader is not to imagine that by those words I anywhere take upon me to define the kind, or the manner of any action, the causes or the physical reason thereof, or that I attribute forces, in a true and physical sense, to certain centres (which are only mathematical points); when at any time I happen to speak of centres as attracting, or as endued with attractive powers.

Scholium

Hitherto I have laid down the definitions of such words as are less known, and explained the sense in which I would have them to be understood in the following discourse. I do not define time, space, place, and motion, as being well known to all. Only I must observe, that the common people conceive those quantities under no other notions but from the relation they bear to sensible objects. And thence arise certain prejudices, for the removing of which it will be convenient to distinguish them into absolute and relative, true and apparent, mathematical and common.

I. Absolute, true, and mathematical time, of itself, and from its own nature, flows equably without relation to anything external, and by another name is called duration: relative, apparent, and common time, is some sensible and external (whether accurate or unequable) measure of duration by the means of motion, which is commonly used instead of true time; such as an hour, a day, a month, a year.

II. Absolute space, in its own nature, without relation to anything external, remains always similar and immovable. Relative space is some movable dimension or measure of the absolute spaces; which our senses determine by its position to bodies; and which is commonly taken for immovable space; such is the dimension of a subterraneous, an aerial, or celestial space, determined by its position in respect of the earth. Absolute and relative space are the same in figure and magnitude; but they do not remain always numerically the same. For if the earth, for instance, moves, a space of our air, which relatively and in respect of the earth remains always the same, will at one time be one part of the absolute space into which the air passes; at another time it will be another part of the same, and so, absolutely understood, it will be continually changed.

III. Place is a part of space which a body takes up, and is according to the space, either absolute or relative. I say, a part of space; not the situation, nor the external surface of the body. For the places of equal solids are always equal; but their surfaces, by reason of their dissimilar figures, are often unequal. Positions properly have no quantity, nor are they so much the places themselves, as the properties of places. The motion of the whole is the same with the sum of the motions of the parts; that is, the translation of the whole, out of its place, is the same thing with the sum of translations of the parts out of their places; and therefore the place of the whole is the same as the sum of the places of the parts, and for that reason, it is internal, and in the whole body.

IV. Absolute motion is the translation of a body from one absolute place into another; and relative motion, the translation from one relative place into another. Thus in a ship under sail, the relative place of a body is that part of the ship which the body possesses; or that part of the cavity which the body fills, and which therefore moves together with the ship: and relative rest is the continuance of the body in the same part of the ship, or of its cavity. But real, absolute rest, is the continuance of the body in the same part of that immovable space, in which the ship itself, its cavity, and all that it contains, is moved. Wherefore, if the earth is really at rest, the body, which relatively rests in the ship, will really and absolutely move with the same velocity which the ship has on the earth. But if the earth also moves, the true and absolute motion of the body will arise, partly from the true motion of the earth, in immovable space, partly from the relative motion of the ship on the earth; and if the body moves also relatively in the ship, its true motion will arise, partly from the true motion of the earth, in immovable space, and partly from the relative motions as well of the ship on the earth, as of the body in the ship; and from these relative motions will arise the relative motion of the body on the earth. As if that part of the earth, where the ship is, was truly moved towards the east, with a velocity of 10010 parts; while the ship itself, with a fresh gale, and full sails, is carried towards the west, with a velocity expressed by 10 of those parts; but a sailor walks in the ship towards the east, with 1 part of the said velocity; then the sailor will be moved truly in immovable space towards the east, with a velocity of 10001 parts, and relatively on the earth towards the west, with a velocity of 9 of those parts.

Absolute time, in astronomy, is distinguished from relative, by the equation or correction of the apparent time. For the natural days are truly unequal, though they are commonly considered as equal, and used for a measure of time; astronomers correct this inequality that they may measure the celestial motions by a more accurate time. It may be, that there is no such thing as an equable motion, whereby time may be accurately measured. All motions may be accelerated and retarded, but the flowing of absolute time is not liable to any change. The duration or perseverance of the existence of things remains the same, whether the motions are swift or slow, or none at all: and therefore this duration ought to be distinguished from what are only sensible measures thereof; and from which we deduce it, by means of the astronomical equation. The necessity of the equation, for determining the times of a phenomenon, is evinced as well from the experiments of the pendulum clock, as by eclipses of the satellites of Jupiter.

As the order of the parts of time is immutable, so also is the order of the parts of space. Suppose those parts to be moved out of their places, and they will be moved (if the expression may be allowed) out of themselves. For times and spaces are, as it were, the places as well of themselves as of all other things. All things are placed in time as to order of succession; and in space as to order of situation. It is from their essence or nature that they are places; and that the primary places of things should be movable, is absurd. These are therefore the absolute places; and translations out of those places, are the only absolute motions.

But because the parts of space cannot be seen, or distinguished from one another by our senses, therefore in their stead we use sensible measures of them. For from the positions and distances of things from any body considered as immovable, we define all places; and then with respect to such places, we estimate all motions, considering bodies as transferred from some of those places into others. And so, instead of absolute places and motions, we use relative ones; and that without any inconvenience in common affairs; but in philosophical disquisitions, we ought to abstract from our senses, and consider things themselves, distinct from what are only sensible measures of them. For it may be that there is no body really at rest, to which the places and motions of others may be referred.

But we may distinguish rest and motion, absolute and relative, one from the other by their properties, causes, and effects. It is a property of rest, that bodies really at rest do rest in respect to one another. And therefore as it is possible, that in the remote regions of the fixed stars, or perhaps far beyond them, there may be some body absolutely at rest; but impossible to know, from the position of bodies to one another in our regions, whether any of these do keep the same position to that remote body, it follows that absolute rest cannot be determined from the position of bodies in our regions.

It is a property of motion, that the parts, which retain given positions to their wholes, do partake of the motions of those wholes. For all the parts of revolving bodies endeavor to recede from the axis of motion; and the impetus of bodies moving forwards arises from the joint impetus of all the parts. Therefore, if surrounding bodies are moved, those that are relatively at rest within them will partake of their motion. Upon which account, the true and absolute motion of a body cannot be determined by the translation of it from those which only seem to rest; for the external bodies ought not only to appear at rest, but to be really at rest. For otherwise, all included bodies, besides their translation from near the surrounding ones, partake likewise of their true motions; and though that translation were not made, they would not be really at rest, but only seem to be so. For the surrounding bodies stand in the like relation to the surrounded as the exterior part of a whole does to the interior, or as the shell does to the kernel; but if the shell moves, the kernel will also move, as being part of the whole, without any removal from near the shell.

A property, near akin to the preceding, is this, that if a place is moved, whatever is placed therein moves along with it; and therefore a body, which is moved from a place in motion, partakes also of the motion of its place. Upon which account, all motions, from places in motion, are no other than parts of entire and absolute motions; and every entire motion is composed of the motion of the body out of its first place, and the motion of this place out of its place; and so on, until we come to some immovable place, as in the before-mentioned example of the sailor. Wherefore, entire and absolute motions can be no otherwise determined than by immovable places; and for that reason I did before refer those absolute motions to immovable places, but relative ones to movable places. Now no other places are immovable but those that, from infinity to infinity, do all retain the same given position one to another; and upon this account must ever remain unmoved; and do thereby constitute immovable space.

The causes by which true and relative motions are distinguished, one from the other, are the forces impressed upon bodies to generate motion. True motion is neither generated nor altered, but by some force impressed upon the body moved; but relative motion may be generated or altered without any force impressed upon the body. For it is sufficient only to impress some force on other bodies with which the former is compared, that by their giving way, that relation may be changed, in which the relative rest or motion of this other body did consist. Again, true motion suffers always some change from any force impressed upon the moving body; but relative motion does not necessarily undergo any change by such forces. For if the same forces are likewise impressed on those other bodies, with which the comparison is made, that the relative position may be preserved, then that condition will be preserved in which the relative motion consists. And therefore any relative motion may be changed when the true motion remains unaltered, and the relative may be preserved when the true suffers some change. Thus, true motion by no means consists in such relations.

The effects which distinguish absolute from relative motion are the forces of receding from the axis of circular motion. For there are no such forces in a circular motion purely relative, but in a true and absolute circular motion, they are greater or less, according to the quantity of the motion. If a vessel, hung by a long cord, is so often turned about that the cord is strongly twisted, then filled with water, and held at rest together with the water; thereupon, by the sudden action of another force, it is whirled about the contrary way, and while the cord is untwisting itself, the vessel continues for some time in this motion; the surface of the water will at first be plain, as before the vessel began to move; but after that, the vessel, by gradually communicating its motion to the water, will make it begin sensibly to revolve, and recede by little and little from the middle, and ascend to the sides of the vessel, forming itself into a concave figure (as I have experienced), and the swifter the motion becomes, the higher will the water rise, till at last, performing its revolutions in the same times with the vessel, it becomes relatively at rest in it. This ascent of the water shows its endeavor to recede from the axis of its motion; and the true and absolute circular motion of the water, which is here directly contrary to the relative, becomes known, and may be measured by this endeavor. At first, when the relative motion of the water in the vessel was greatest, it produced no endeavor to recede from the axis; the water showed no tendency to the circumference, nor any ascent towards the sides of the vessel, but remained of a plain surface, and therefore its true circular motion had not yet begun. But afterwards, when the relative motion of the water had decreased, the ascent thereof towards the sides of the vessel proved its endeavor to recede from the axis; and this endeavor showed the real circular motion of the water continually increasing, till it had acquired its greatest quantity, when the water rested relatively in the vessel. And therefore this endeavor does not depend upon any translation of the water in respect of the ambient bodies, nor can true circular motion be defined by such translation. There is only one real circular motion of any one revolving body, corresponding to only one power of endeavoring to recede from its axis of motion, as its proper and adequate effect; but relative motions, in one and the same body, are innumerable, according to the various relations it bears to external bodies, and, like other relations, are altogether destitute of any real effect, any otherwise than they may perhaps partake of that one only true motion. And therefore in their system who suppose that our heavens, revolving below the sphere of the fixed stars, carry the planets along with them; the several parts of those heavens, and the planets, which are indeed relatively at rest in their heavens, do yet really move. For they change their position one to another (which never happens to bodies truly at rest), and being carried together with their heavens, partake of their motions, and as parts of revolving wholes, endeavor to recede from the axis of their motions.

Wherefore relative quantities are not the quantities themselves, whose names they bear, but those sensible measures of them (either accurate or inaccurate), which are commonly used instead of the measured quantities themselves. And if the meaning of words is to be determined by their use, then by the names time, space, place, and motion, their [sensible] measures are properly to be understood; and the expression will be unusual, and purely mathematical, if the measured quantities themselves are meant. On this account, those violate the accuracy of language, which ought to be kept precise, who interpret these words for the measured quantities. Nor do those less defile the purity of mathematical and philosophical truths, who confound real quantities with their relations and sensible measures.

It is indeed a matter of great difficulty to discover, and effectually to distinguish, the true motions of particular bodies from the apparent; because the parts of that immovable space, in which those motions are performed, do by no means come under the observation of our senses. Yet the thing is not altogether desperate; for we have some arguments to guide us, partly from the apparent motions, which are the differences of the true motions; partly from the forces, which are the causes and effects of the true motions. For instance, if two globes, kept at a given distance one from the other by means of a cord that connects them, were revolved about their common centre of gravity, we might, from the tension of the cord, discover the endeavor of the globes to recede from the axis of their motion, and from thence we might compute the quantity of their circular motions. And then if any equal forces should be impressed at once on the alternate faces of the globes to augment or diminish their circular motions, from the increase or decrease of the tension of the cord, we might infer the increment or decrement of their motions; and thence would be found on what faces those forces ought to be impressed, that the motions of the globes might be most augmented; that is, we might discover their hindmost faces, or those which, in the circular motion, do follow. But the faces which follow being known, and consequently the opposite ones that precede, we should likewise know the determination of their motions. And thus we might find both the quantity and the determination of this circular motion, even in an immense vacuum, where there was nothing external or sensible with which the globes could be compared. But now, if in that space some remote bodies were placed that kept always a given position one to another, as the fixed stars do in our regions, we could not indeed determine from the relative translation of the globes among those bodies, whether the motion did belong to the globes or to the bodies. But if we observed the cord, and found that its tension was that very tension which the motions of the globes required, we might conclude the motion to be in the globes, and the bodies to be at rest; and then, lastly, from the translation of the globes among the bodies, we should find the determination of their motions. But how we are to obtain the true motions from their causes, effects, and apparent differences, and the converse, shall be explained more at large in the following treatise. For to this end it was that I composed it.

AXIOMS, OR
LAWS OF MOTION

Law I

Every body continues in its state of rest, or of uniform motion in a right line, unless it is compelled to change that state by forces impressed upon it.

Projectiles continue in their motions, so far as they are not retarded by the resistance of the air, or impelled downwards by the force of gravity. A top, whose parts by their cohesion are continually drawn aside from rectilinear motions, does not cease its rotation, otherwise than as it is retarded by the air. The greater bodies of the planets and comets, meeting with less resistance in freer spaces, preserve their motions both progressive and circular for a much longer time.

Law II

The change of motion is proportional to the motive force impressed; and is made in the direction of the right line in which that force is impressed.

If any force generates a motion, a double force will generate double the motion, a triple force triple the motion, whether that force be impressed altogether and at once, or gradually and successively. And this motion (being always directed the same way with the generating force), if the body moved before, is added to or subtracted from the former motion, according as they directly conspire with or are directly contrary to each other; or obliquely joined, when they are oblique, so as to produce a new motion compounded from the determination of both.

Law III

To every action there is always opposed an equal reaction: or, the mutual actions of two bodies upon each other are always equal, and directed to contrary parts.

Whatever draws or presses another is as much drawn or pressed by that other. If you press a stone with your finger, the finger is also pressed by the stone. If a horse draws a stone tied to a rope, the horse (if I may so say) will be equally drawn back towards the stone; for the distended rope, by the same endeavor to relax or unbend itself, will draw the horse as much towards the stone as it does the stone towards the horse, and will obstruct the progress of the one as much as it advances that of the other. If a body impinge upon another, and by its force change the motion of the other, that body also (because of the equality of the mutual pressure) will undergo an equal change, in its own motion, towards the contrary part. The changes made by these actions are equal, not in the velocities but in the motions of bodies; that is to say, if the bodies are not hindered by any other impediments. For, because the motions are equally changed, the changes of the velocities made towards contrary parts are inversely proportional to the bodies. This law takes place also in attractions, as will be proved in the next Scholium.

Corollary I

A body, acted on by two forces simultaneously, will describe the diagonal of a parallelogram in the same time as it would describe the sides by those forces separately.

If a body [Fig. 1] in a given time, by the force M impressed apart in the place A, should with an uniform motion be carried

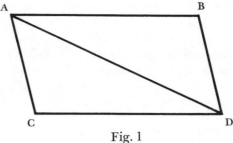

Fig. 1

from A to B, and by the force N impressed apart in the same place, should be carried from A to C, let the parallelogram ABCD be completed, and, by both forces acting together, it will in the same time be carried in the diagonal from A to D. For since the force N acts in the direction of the line AC, parallel to BD, this force (by the second Law) will not at all alter the velocity generated by the other force M, by which the body is carried towards the line BD. The body therefore will arrive at the line BD in the same time, whether the force N be impressed or not; and therefore at the end of that time it will be found somewhere in the line BD. By the same argument, at the end of the same time it will be found somewhere in the line CD. Therefore it will be found in the point D, where both lines meet. But it will move in a right line from A to D, by Law I.

.   .   .   .   .

Corollary III

The quantity of motion, which is obtained by taking the sum of the motions directed towards the same parts, and the difference of those that are directed to contrary parts, suffers no change from the action of bodies among themselves.

For action and its opposite reaction are equal, by Law III, and therefore, by Law II, they produce in the motions equal changes towards opposite parts. Therefore if the motions are directed towards the same parts, whatever is added to the motion of the preceding body will be subtracted from the motion of that which follows; so that the sum will be the same as before. If the bodies meet, with contrary motions, there will be an equal deduction from the motions of both; and therefore the difference of the motions directed towards opposite parts will remain the same.

Thus, if a spherical body A is 3 times greater than the spherical body B, and has a velocity $=2$, and B follows in the same direction with a velocity $=10$, then the

$$\text{motion of A : motion of B} = 6:10.$$

Suppose, then, their motions to be of 6 parts and of 10 parts, and the sum will be 16 parts. Therefore, upon the meeting of the bodies, if A acquire 3, 4, or 5 parts of motion, B will lose as many; and therefore after reflection A will proceed with 9, 10, or 11 parts, and B with 7, 6, or 5 parts; the sum remaining always of 16 parts as before. If the body A acquire 9, 10, 11, or 12 parts of motion, and therefore after meeting proceed with 15, 16, 17, or 18 parts, the body B, losing so many parts as A has got, will either proceed with 1 part, having lost 9, or stop and remain at rest, as having lost its whole progressive motion of 10 parts; or it will go back with 1 part, having not only lost its whole motion, but (if I may so say) one part more; or it will go back with 2 parts, because a progressive motion of 12 parts is taken off. And so the sums of the conspiring motions,

$$15+1 \quad \text{or} \quad 16+0,$$

and the differences of the contrary motions,

$$17-1 \quad \text{and} \quad 18-2,$$

will always be equal to 16 parts, as they were before the meeting and reflection of the bodies. But the motions being known with which the bodies proceed after reflection, the velocity of either will be also known, by taking the velocity after to the velocity before reflection, as the motion after is to the motion before. As in the last case, where the

motion of A before reflection (6) : motion of A after (18)
    = velocity of A before (2) : velocity of A after (x);

that is,

$$6:18 = 2:x, x = 6.$$

But if the bodies are either not spherical, or, moving in different right lines, impinge obliquely one upon the other, and their motions after reflection are required, in those cases we are first to determine the position of the plane that touches the bodies in the point of impact, then the motion of each body (by Cor. II) is to be resolved into two, one perpendicular to that plane, and the other parallel to it. This done, because the bodies act upon each other in the direction of a line perpendicular to this plane, the parallel motions are to be retained the same after reflection as before; and to the perpendicular motions we are to assign equal changes towards the contrary parts; in such manner that the sum of the conspiring and the difference of the contrary motions may remain the same as before. From such kind of reflections sometimes arise also the circular motions of bodies about their own centres. But these are cases which I do not consider in what follows; and it would be too tedious to demonstrate every particular case that relates to this subject.

Corollary IV

The common centre of gravity of two or more bodies does not alter its state of motion or rest by the actions of the bodies among themselves; and therefore the common centre of gravity of all bodies acting upon each other (excluding external actions and impediments) is either at rest, or moves uniformly in a right line.

For if two points proceed with an uniform motion in right lines, and their distance be divided in a given ratio, the dividing point will be either at rest, or proceed uniformly in a right line. This is demonstrated hereafter in Lem. XXIII and Corollary, when the points are moved in the same plane; and by a like way of arguing, it may be demonstrated when the points are not moved in the same plane. Therefore if any number of bodies move uniformly in right lines, the common centre of gravity of any two of them is either at rest, or proceeds uniformly in a right line; because the line which connects the centres of those two bodies so moving is divided at that common centre in a given ratio. In like manner the common centre of those two and that of a third body will be either at rest or moving uniformly in a right line; because at that centre the distance between the common centre of the two bodies, and the centre of this last, is divided in a given ratio. In like manner the common centre of these three, and of a fourth body, is either at rest, or moves uniformly in a right line; because the distance between the common centre of the three bodies, and the centre of the fourth, is there also divided in a given ratio, and so on *in infinitum.* Therefore, in a system of bodies where there is neither any mutual action among themselves, nor any foreign force impressed upon them from without, and which consequently move uniformly in right lines, the common centre of gravity of them all is either at rest or moves uniformly forwards in a right line.

Moreover, in a system of two bodies acting upon each other, since the distances between their centres and the common centre of gravity of both are reciprocally as the bodies, the relative motions of those bodies, whether of approaching to or of receding from that centre, will be equal among themselves. Therefore since the changes which happen to motions are equal and directed to contrary parts, the common centre of those bodies, by their mutual action between themselves, is neither accelerated nor retarded, nor suffers any change as to its state of motion or rest. But in a system of several bodies, because the common centre of gravity of any two acting upon each other suffers no change in its state by that action; and much less the common centre of gravity of the others with which that action does not intervene; but the distance between those two centres is divided by the common centre of gravity of all the bodies into parts inversely proportional to the total sums of those bodies whose centres they are; and therefore while those two centres retain their state of motion or rest, the common centre of all does also retain its state: it is manifest that the common centre of all never suffers any change in the state of its motion or rest from the actions of any two bodies between themselves. But in such a system all the actions of the bodies among themselves either happen between two bodies, or are composed of actions interchanged between some two bodies; and therefore they do never produce any alteration in the common centre of all as to its state of motion or rest. Wherefore since that centre, when the bodies do not act one upon another, either is at rest or moves uniformly forwards in some right line, it will, notwithstanding the mutual actions of the bodies among themselves, always continue in its state, either of rest, or of proceeding uniformly in a right line, unless it is forced out of this state by the action of some power impressed from without upon the whole system. And therefore the same law takes place in a system consisting of many bodies as in one single body, with regard to their persevering in their state of motion or of rest. For the progressive motion, whether of one single body, or of a whole system of bodies, is always to be estimated from the motion of the centre of gravity.

Corollary V

The motions of bodies included in a given space are the same among themselves, whether that space is at rest, or moves uniformly forwards in a right line without any circular motion.

For the differences of the motions tending towards the same parts, and the sums of those that tend towards contrary parts, are, at first (by supposition), in both cases the same; and it is from those sums and differences that the collisions and impulses do arise with which the bodies impinge one upon another. Wherefore (by Law II), the effects of those collisions will be equal in both cases; and therefore the mutual motions of the bodies among themselves in the one case will remain equal to the motions of the bodies among themselves in the other. A clear proof of this we have from the experiment of a ship; where all motions happen after the same manner whether the ship is at rest, or is carried uniformly forwards in a right line.

Corollary VI

If bodies, moved in any manner among themselves, are urged in the direction of parallel lines by equal accelerative forces, they will all continue to move among themselves, after the same manner as if they had not been urged by those forces.

For these forces acting equally (with respect to the quantities of the bodies to be moved), and in the direction of parallel lines, will (by Law II) move all the bodies equally (as to velocity), and therefore will never produce any change in the positions or motions of the bodies among themselves.

Absolute and Relational Theories of Space and Time

LEIBNIZ and CLARKE, from
The Correspondence

The correspondence between Gottfried Wilhelm von Leibniz (1646–1716) and Samuel Clarke (1675–1729) consists of five letters which were written between 1715–1716 to the Hanoverian Princess Caroline. Leibniz's view of space and time was one major alternative to Newton's. According to Newton, bodies are located by specifying that they occupy a certain point or points of Absolute Space. But Leibniz argued that Newton's concept of space was unintelligible. Clarke, who defended Newton's theory, asserted at various times during the correspondence that Absolute Space is a substance, an infinite being, and several times he declared it was a property. These were not simply Samuel Clarke's views of Absolute Space; it is now reasonably certain that Clarke wrote his replies with the approval and perhaps the editorial benefit of Newton.

Leibniz's replies are to the effect that space cannot be an eternal, infinite substance, because there is really only one such entity, namely, God, so that space would have to be identical with God, which is false. Neither can space be a property. For, Leibniz says, the Newtonians believe that there can be space with no bodies in it. If space is a property, then like all properties it is a property of something. But if space were devoid of bodies, then aside from space there is nothing of which it is a property, and this is absurd.

Perhaps the most striking set of criticisms that Leibniz made were those which rested upon his principle of indiscernibles. According to this principle there cannot be two things which have all their properties in common. Leibniz applied his principle to Absolute Space and Absolute Time. Absolute Space, for example, is supposed to consist of parts which are completely similar to each other. In a parallel way, any two instants or moments of time are supposed to be completely alike. Leibniz concluded that Absolute Space cannot have indistinguishable parts, nor can Absolute Time have indistinguishable moments or instants. This conclusion is disastrous for the Newtonian position and we leave the reader to decide how satisfactory is Clarke's reply.

The reader will find still other criticisms raised by Leibniz which have a very modern ring to them (though they are not the fairest criticisms in the world). For example, according to Newton, a body could be either at rest or in uniform motion in a straight line, but it would be impossible to determine which is the case by using the laws of mechanics. Leibniz argued that if there is motion, then it is possible for it to be observed. "When there is no change which can be observed, then there is no change." This is a criticism which we have come to regard as typical of modern positivism, not of rationalist metaphysics. On the other hand, as the reader will see, Leibniz relied heavily upon the actions of God to explain certain features which he thought described the world—such as the absence of a vacuum.

Leibniz did more than criticize the conceptual apparatus of Newton's theory of motion. He offered a fresh analysis of the concepts of space and time, arguing for a relational theory of space and time rather than their Absolute counterparts. According to Leibniz, space, for example, is the order or perhaps the relative arrangement of coexistent bodies. It is apparent from this characterization of space that if there are no bodies, there can be no space. Time, too, would not exist if there were no bodies, for Leibniz asserted that time is the order of succession of coexistent bodies. Clarke, for his part, raised a host of interesting objections to Leibniz's relational theory, which pointed out the need for more that this rough description of space and time. Clarke's assertion that there is empirical evidence for motions with respect to Absolute Space elicited only a faint reaction from Leibniz. Perhaps he believed that Newtonian space and time were so patently unsound that any discussion of empirical evidence for their existence was science-fiction.

Leibniz's view of space and time was not widely accepted, and the Newtonian view soon became standard until at least the middle of the nineteenth century. The principal reason for this fact is probably that it was not enough to offer trenchant criticisms of Newton's concepts of space and time to gain acceptance of one's own view. The Newtonian concepts figured in an explanatory scheme which was enormously successful, so that nothing short of an alternative system of mechanics incorporating Leibniz's theory of space and time could have turned that relational theory into a serious alternative to the Theory of Absolute Space and Time. There are indications that Leibniz had already made a start on the new mechanics, but his results seem too fragmentary to constitute an alternative to Newtonian mechanics.

NOTE: In *The Correspondence,* the number of a paragraph usually introduces the rejoinder to a similarly numbered section in the preceding letter.

MR. LEIBNIZ'S THIRD PAPER[1]

Being
An Answer to Dr. Clarke's Second Reply

3. These gentlemen maintain therefore, that space is real absolute being. But this involves them in great difficulties; for such a being must needs be eternal and infinite. Hence some have believed it to be God himself, or, one of his attributes, his immensity. But since space consists of parts, it is not a thing which can belong to God.

4. As for my own opinion, I have said more than once, that I hold space to be something merely[2] relative, as time is; that I hold it to be an order of coexistences, as time is an order of successions. For space denotes, in terms of possibility, an order of things which exist at the same time, considered as existing together; without enquiring into their manner of existing. And when many things are seen together, one perceives that order of things among themselves.

5. I have many demonstrations, to confute the fancy of those who take space to be a substance, or at least an absolute being. But I shall only use, at the present, one demonstration, which the author here gives me occasion to insist upon. I say then, that if space was an absolute being, there would something happen for which it would be impossible there should be a suffcent reason. Which is against my axiom. And I prove it thus. Space is something absolutely uniform; and, without the things placed in it, one point of space does not absolutely differ in any respect whatsoever from another point of space. Now from hence it follows, (supposing space to be something in itself, besides the order of

bodies among themselves,) that 'tis impossible there should be a reason, why God, preserving the same situations of bodies among themselves, should have placed them in space after one certain particular manner, and not otherwise; why every thing was not placed the quite contrary way, for instance, by changing East into West. But if space is nothing else, but that order or relation, and is nothing at all without bodies, but the possibility of placing them, then those two states, the one such as it now is, the other supposed to be the quite contrary way, would not at all differ from one another. Their difference therefore is only to be found in our chimerical supposition of the reality of space in itself. But in truth the one would exactly be the same thing as the other, they being absolutely indiscernible, and consequently there is no room to enquire after a reason of the preference of the one to the other.

6. The case is the same with respect to time. Supposing any one should ask, why God did not create every thing a year sooner; and the same person should infer from thence, that God has done something, concerning which 'tis not possible there should be a reason, why he did it so, and not otherwise: the answer is, that his inference would be right, if time was any thing distinct from things existing in time. For it would be impossible there should be any reason, why things should be applied to such particular instants, rather than to others, their succession continuing the same. But then the same argument proves, that instants, consider'd without the things, are nothing at all; and that they consist only in the successive order of things: which order remaining the same, one of the two states, viz. that of a supposed anticipation, would not at all differ, nor could be discerned from, the other which now is.

DR. CLARKE'S THIRD REPLY[3]

2. Undoubtedly nothing is, without a sufficent reason why it is, rather than not; and why it is thus, rather than otherwise. But in things in their own nature indifferent, mere will, without any thing external to influence it, is alone that sufficient reason. As in the instance of God's creating or placing any particle of matter in one place rather than in another, when all places are originally alike. And the case is the same, even though space were nothing real, but only the mere order of bodies: for still it would be absolutely indifferent, and there could be no other reason but

mere will, why three equal particles should be placed or ranged in the order *a, b, c,* rather than in the contrary order. And therefore no argument can be drawn from this indifferency of all places, to prove that no space is real. For different spaces are really different or distinct one from another, though they be perfectly alike. And there is this evident absurdity in supposing space not to be real, but to be merely the order of bodies; that, according to that notion, if the earth and sun and moon had been placed where the remotest fixed stars now are, (provided they were placed in the same order and distance they now are with regard one to another), it would not only have been, (as this learned author rightly says), *la même chose,* the same thing in effect; which is very true: but it would also follow, that they would then have been in the same place too, as they are now: which is an express contradiction.

The ancients[4] did not call all space which is void of bodies, but only extramundane space, by the name of imaginary space. The meaning of which, is not, that such space is not real;[5] but only that we are wholly ignorant what kinds of things are in that space. Those writers, who by the word, *imaginary,* meant at any time to affirm that space was not real; did not thereby prove, that it was not real.

3. Space is not a being, an eternal and infinite being, but a property, or a consequence of the existence of a being infinite and eternal. Infinite space, is immensity: but immensity is not God: and therefore infinite space, is not God. Nor is there any difficulty in what is here alleged about space having parts. For infinite space is one, absolutely and essentially indivisible: and to suppose it parted, is a contradiction in terms; because there must be space in the partition itself; which is to suppose it parted, and yet not parted at the same time.[6] The immensity of God, is no more a dividing of his substance into parts; than his duration, or continuance of existing, is a dividing of his existence into parts. There is no difficulty here, but what arises from the figurative abuse of the word, *parts.*

4. If space was nothing but the order of things coexisting, it would follow, that if God should remove in a straight line the whole material world entire, with any swiftness whatsoever; yet it would still always continue in the same place; and that nothing would receive any shock upon the most sudden stopping of that motion. And if time was nothing but the order of succession of

created things; it would follow, that if God had created the world millions of ages sooner than he did, yet it would not have been created at all the sooner. Further: space and time are quantities; which situation and order are not.

5. The argument in this paragraph, is; that because space is uniform or alike, and one part does not differ from another, therefore the bodies created in one place, if they had been created in another place, (supposing them to keep the same situation with regard to each other,) would still have been created in the same place as before: which is a manifest contradiction. The uniformity of space, does indeed prove, that there could be no (external) reason why God should create things in one place rather than in another: but does that hinder his own will, from being to itself a sufficient reason of acting in any place, when all places are indifferent or alike, and there be good reason to act in some place?

MR. LEIBNIZ'S FOURTH PAPER[7]

Being
An Answer to Dr. Clarke's Third Reply

3. 'Tis a thing indifferent, to place three bodies, equal and perfectly alike, in any order whatsoever; and consequently they will never be placed in any order, by him who does nothing without wisdom. But then he being the author of things, no such things will be produced by him at all; and consequently there are no such things in nature.

4. There is no such thing as two individuals indiscernible from each other. An ingenious gentleman of my acquaintance, discoursing with me, in the presence of Her Electoral Highness the Princess Sophia,[8] in the garden of Herrenhausen; thought he could find two leaves perfectly alike. The Princess defied him to do it, and he ran all over the garden a long time to look for some; but it was to no purpose. Two drops of water, or milk, viewed with a microscope, will appear distinguishable from each other. This is an argument against atoms; which are confuted as well as a vacuum, by the principles of true metaphysics.

5. Those great principles of a *sufficient reason,* and of the *identity of indiscernibles,* change the state of metaphysics. That science becomes real and demonstrative by means of these principles; whereas before, it did generally consist in empty words.

6. To suppose two things indiscernible, is to suppose the same thing under two names. And therefore to suppose that the universe could have had at first another position of time and place, than that which it actually had, and yet that all the parts of the universe should have had the same situation among themselves, as that which they actually had—such a supposition, I say, is an impossible fiction.

. . . . .

8. If space is a property or attribute, it must be the property of some substance. But what substance will that bounded empty space be an affection or property of, which the persons I am arguing with, suppose to be between two bodies?

9. If infinite space is immensity, finite space will be the opposite to immensity, that is, 'twill be mensurability, or limited extension. Now extension must be the affection of some thing extended. But if that space be empty, it will be an attribute without a subject, an extension without any thing extended. Wherefore by making space a property, the author falls in with my opinion, which makes it an order of things, and not any thing absolute.

10. If space is an absolute reality; far from being a property or an accident opposed to substance, it will have a greater reality than substances themselves.[9] God cannot destroy it, nor even change it in any respect. It will be not only immense in the whole, but also immutable and eternal in every part. There will be an infinite number of eternal things besides God.

. . . . .

13. To say that God can cause the whole universe to move forward in a right line, or in any other line, without making otherwise any alteration in it; is another chimerical supposition. For, two states indiscernible from each other, are the same state; and consequently, 'tis a change without any change. Besides, there is neither rhyme nor reason in it. But God does nothing without reason; and 'tis impossible there should be any here. Besides, it would be *agendo nihil agere,* as I have just now said, because of the indiscernibility.

14. These are *idola tribus,* mere chimeras, and superficial imaginations. All this is only grounded upon the supposition, that imaginary space is real.

15. It is a like fiction, (that is) an impossible one, to suppose that God might have created the world some millions of years sooner. They who run into such kind of fictions, can give no answer to one that should argue for the eternity of the world. For since God does nothing without reason, and no reason can be given why he did not create the world sooner; it will follow, either that he has created nothing at all, or that he created the world before any assignable time, that is, that the world is eternal. But when once it has been shown, that the beginning, whenever it was, is always the same thing; the question, why it was not otherwise ordered, becomes needless and insignificant.

16. If space and time were any thing absolute, that is, if they were any thing else, besides certain orders of things; then indeed my assertion would be a contradiction. But since it is not so, the hypothesis [that space and time are any thing absolute][10] is contradictory, that is, 'tis an impossible fiction.

17. And the case is the same as in geometry; where by the very supposition that a figure is greater than it really is, we sometimes prove that it is not greater. This indeed is a contradiction; but it lies in the hypothesis, which appears to be false for that very reason.

18. Space being uniform, there can be neither any external nor internal reason, by which to distinguish its parts, and to make any choice among them. For, any external reason to discern between them, can only be grounded upon some internal one. Otherwise we should discern what is indiscernible, or choose without discerning. A will without reason, would be the chance of the Epicureans.[11] A God, who should act by such a will, would be a God only in name. The cause of these errors proceeds from want of care to avoid what derogates from the divine perfections.

19. When two things which cannot both be together, are equally good; and neither in themselves, nor by their combination with other things, has the one any advantage over the other; God will produce neither of them.

.   .   .   .   .

21. . . . In like manner, to admit a vacuum in nature, is ascribing to God a very imperfect work: 'tis violating the grand principle of the necessity of a sufficent reason; which many have talked of, without understanding its true meaning; as I have lately

shown, in proving, by that principle, that space is only an order of things, as time also is, and not at all an absolute being. To omit many other arguments against a vacuum and atoms, I shall here mention those which I ground upon God's perfection, and upon the necessity of a sufficient reason. I lay it down as a principle, that every perfection, which God could impart to things without derogating from their other perfections, has actually been imparted to them. Now let us fancy a space wholly empty. God could have placed some matter in it, without derogating in any respect from all other things: therefore he has actually placed some matter in that space: therefore, there is no space wholly empty; therefore all is full. The same argument proves that there is no corpuscle, but what is subdivided.

DR. CLARKE'S FOURTH REPLY[12]

3 and 4. This argument, if it was true, would prove that God neither has created, nor can possibly create any matter at all. For the perfectly solid parts of all matter, if you take them of equal figure and dimensions (which is always possible in supposition,) are exactly alike; and therefore it would be perfectly indifferent if they were transposed in place; and consequently it was impossible (according to this learned author's argument,) for God to place them in those places wherein he did actually place them at the creation, because he might as easily have transposed their situation. 'Tis very true, that no two leaves, and perhaps no two drops of water are exactly alike; becouse they are bodies very much compounded. But the case is very different in the parts of simple solid matter. And even in compounds, there is no impossibility for God to make two drops of water exactly alike. And if he should make them exactly alike, yet they would never the more become one and the same drop of water, because they were alike. Nor would the place of the one, be the place of the other; though it was absolutely indifferent, which was placed in which place. The same reasoning holds likewise concerning the original determination of motion, this way or the contrary way.

5 and 6. Two things, by being exactly alike, do not cease to be two. The parts of time, are as exactly like to each other, as those of space: yet two points of time, are not the same point of time, nor are they two names of only the same point of time. Had God

created the world but this moment, it would not have been created at the time it was created. And if God has made (or can make) matter finite in dimensions, the material universe must consequently be in its nature moveable; for nothing that is finite, is immoveable. To say therefore that God could not have altered the time or place of the existence of matter, is making matter to be necessarily infinite and eternal, and reducing all things to necessity and fate.

·   ·   ·   ·   ·

8. Space void of body, is the property of an incorporeal substance. Space is not bounded by bodies, but exists equally within and without bodies. Space is not inclosed between bodies; but bodies, existing in unbounded space, are, themselves only, terminated by their own dimensions.

9. Void space, is not an attribute without a subject; because, by void space, we never mean space void of every thing, but void of body only. In all void space, God is certainly present, and possibly many other substances which are not matter, being neither tangible, nor objects of any of our senses.

10. Space is not a substance, but a property;[13] and if it be a property of that which is necessary, it will consequently (as all other properties of that which is necessary must do,) exist more necessarily, (thought it be not itself a substance,) than those substances themselves which are not necessary. Space is immense, and immutable, and eternal; and so also is duration. Yet it does not at all from hence follow, that any thing is eternal *hors de Dieu*. For space and duration are not *hors de Dieu,* but are caused by, and are immediate and necessary consequences of his existence.[14] And without them, his eternity and ubiquity (or omnipresence) would be taken away.

·   ·   ·   ·   ·

13. If the world be finite in dimensions, it is moveable by the power of God and therefore my argument drawn from that moveableness is conclusive. Two places, though exactly alike, are not the same place. Nor is the motion or rest of the universe, the same state; any more than the motion or rest of a ship, is the same state, because a man shut up in the cabin cannot perceive whether

the ship sails or not, so long as it moves uniformly. The motion of the ship, though the man perceives it not, is a real different state, and has real different effects; and, upon a sudden stop, it would have other real effects; and so likewise would an indiscernible motion of the universe. To this argument, no answer has ever been given. It is largely insisted on by Sir Isaac Newton in his *Mathematical Principles,* (Definit. 8.) where, from the consideration of the properties, causes, and effects of motion, he shows the difference between real motion, or a body's being carried from one part of space to another; and relative motion, which is merely a change of the order or situation of bodies with respect to each other. This argument is a mathematical one; showing, from real effects, that there may be real motion where there is none relative; and relative motion, where there is none real: and is not to be answered, by barely asserting the contrary.

14. The reality of space is not a supposition, but is proved by the foregoing arguments, to which no answer has been given. Nor is any answer given to that other argument, that space and time are quantities, which situation and order are not.

15. It was no impossibility for God to make the world sooner or later than he did: nor is it at all impossible for him to destroy it sooner or later than it shall actually be destroyed. As to the notion of the world's eternity; they who suppose matter and space to be the same, must indeed suppose the world to be not only infinite and eternal, but necessarily so; even as necessarily as space and duration, which depend not on the will, but on the existence of God. But they who believe that God created matter in what quantity, and at what particular time, and in what particular spaces he pleased, are here under no difficulty. For the wisdom of God may have very good reasons for creating this world, at that particular time he did; and may have made other kinds of things before this material world began, and may make other kinds of things after the world is destroyed.

.   .   .   .   .

19. This argument, (as I now observed, §3) if it proves anything, proves that God neither did nor can create any matter at all; because the situation of equal and similar parts of matter, could not but be originally indifferent: as was also the original determination of their motions, this way, or the contrary way.

MR. LEIBNIZ'S FIFTH PAPER

.   .   .   .   .

34. The author objects against me the vacuum discovered by Mr. Guerike[15] of Magdeburg, which is made by pumping the air out of a receiver; and he pretends that there is truly a perfect vacuum, or a space without matter, (at least in part,) in that receiver. The Aristotelians and Cartesians, who do not admit a true vacuum, have said in answer to that experiment of Mr. Guerike, as well as to that of Torricellius[16] of Florence, (who emptied the air out of a glass-tube by the help of quicksilver), that there is no vacuum at all in the tube or in the receiver; since glass has small pores, which the beams of light, the effluvia of the lodestone, and other very thin fluids may go through. I am of their opinion: and I think the receiver may be compared to a box full of holes in the water, having fish or other gross bodies shut up in it; which being taken out, their place would nevertheless be filled up with water. There is only this difference; that though water be fluid and more yielding than those gross bodies, yet it is as heavy and massive, if not more, than they: whereas the matter which gets into the receiver in the room of the air is much more subtile. The new sticklers for a vacuum allege in answer to this instance, that it is not the grossness of matter, but its mere quantity, that makes resistance; and consequently that there is of necessity more vacuum, where there is less resistance. They add, that the subtleness of matter has nothing to do here; and that the particles of quicksilver are as subtle and fine as those of water; and yet that quicksilver resists about ten times more. To this I reply, that it is not so much the quantity of matter, as its difficulty of giving place, that makes resistance. For instance, floating timber contains less of heavy matter, than an equal bulk of water does; and yet it makes more resistance to a boat, than the water does.

.   .   .   .   .

46. It appears that the author confounds immensity, or the extension of things, with the space according to which that extension is taken. Infinite space, is not the immensity of God; finite space, is not the extension of bodies: as time is not their

SPACE, TIME AND MOTION: *Leibniz and Clarke*   99

duration. Things keep their extension; but they do not always keep their space. Every thing has it own extension, its own duration; but it has not its own time, and does not keep its own space.

47. I will here show, how men come to form to themselves the notion of space. They consider that many things exist at once and they observe in them a certain order of co-existence, according to which the relation of one thing to another is more or less simple. This order, is their *situation* or distance. When it happens that one of those co-existent things changes its relation to a multitude of others, which do not change their relation among themselves; and that another thing, newly come, acquires the same relation to the others, as the former had; we then say, it is come into the place of the former; and this change, we call a motion in that body, wherein is the immediate cause of the change. And though many, or even all the co-existent things, should change according to certain known rules of direction and swiftness; yet one may always determine the relation of situation, which every co-existent acquires with respect to every other co-existent; and even that relation which any other co-existent would have to this, or which this would have to any other, if it had not changed, or if it had changed any otherwise. And supposing, or feigning, that among those co-existents, there is a sufficient number of them, which have undergone no change, then we may say, that those which have such a relation to those fixed existents, as others had to them before, have now the *same place* which those others had. And that which comprehends all those places, is called *space*. Which shows, that in order to have an idea of place, and consequently of space, it is sufficient to consider these relations, and the rules of their changes, without needing to fancy any absolute reality out of the things whose situation we consider. And, to give a kind of a definition: *place* is that, which we say is the same to A and, to B, when the relation of the co-existence of B, with C, E, F, G, etc. agrees perfectly with the relation of the co-existence, which A had with the same C, E, F, G, etc. supposing there has been no cause of change in C, E, F, G, etc. It may be said also, without entering into any further particularity, that *place* is that, which is the same in different moments to different existent things, when their relations of co-existence with certain other existents, which are supposed to con-

tinue fixed from one of those moments to the other, agree entirely together. And *fixed existents* are those, in which there has been no cause of any change of the order of their co-existence with others; or (which is the same thing,) in which there has been no motion. Lastly, *space* is that, which results from places taken together. And here it may not be amiss to consider the difference between place, and the relation of situation, which is in the body that fills up the place. For, the place of A and B, is the same, whereas the relation of A to fixed bodies, is not precisely and individually the same, as the relation which B (that comes into its place) will have to do [with] the same fixed bodies; but these relations agree only. For, two different subjects, as A and B, cannot have precisely the same individual affection; it being impossible, that the same individual accident should be in two subjects, or pass from one subject to another. But the mind not contented with an agreement, looks for an identity, for something that should be truly the same; and conceives it as being extrinsic to the subjects: and this is what we call *place* and *space*. But this can only be an ideal thing; containing a certain order, wherein the mind conceives the application of relations. In like manner, as the mind can fancy to itself an order made up of genealogical lines, whose bigness would consist only in the number of generations wherein every person would have his place: and if to this one should add the fiction of a *metempsychosis,* and bring in the same human souls again; the persons in those lines might change place; he who was a father, or a grandfather, might become a son, or a grandson, etc. And yet those genealogical places, lines, and spaces, though they should express real truth, would only be ideal things. I shall allege another example, to show how the mind uses, upon occasion of accidents which are in subjects, to fancy to itself something answerable to those accidents, out of the subjects. The ratio or proportion between two lines L and M, may be conceived three several ways: as a ratio of the greater L, to the lesser M; as a ratio of the lesser M, to the greater L; and lastly, as something abstracted from both, that is, as the ratio between L and M, without considering which is the antecedent, or which the consequent; which the subject, and which the object. And thus it is, that proportions are considered in music. In the first way of considering them, L the greater; in the second, M the lesser, is the subject of that accident, which philosophers call relation. But, which of them will be the subject, in the third way of considering them? It cannot be said that both of them, L and M together, are the subject of such an accident; for if so, we should have an accident in two subjects, with one leg in one, and the other in the other; which is contrary to the notion of accidents. Therefore we must say, that this relation, in this third way of considering it, is indeed out of the subjects; but being neither a substance, nor an accident, it must be a mere ideal thing, the consideration of which is nevertheless useful. To conclude:[17] I have here done much like Euclid, who not being able to make his readers well understand what *ratio* is absolutely in the sense of geometricians; defines what are the *same ratios*. Thus, in like manner, in order to explain what *place* is, I have been content to define what is the *same place*. Lastly; I observe, that the traces of moveable bodies, which they leave sometimes upon the immoveable ones on which they are moved, have given men occasion to form in their imagination such an idea, as if some trace did still remain, even when there is nothing unmoved. But this is a mere ideal thing, and imports only, that if there was any unmoved thing there, the trace might be marked out upon it. And 'tis this analogy, which makes men fancy places, traces and spaces; though those things consist only in the truth of relations, and not at all in any absolute reality.

.   .   .   .   .

48. To conclude. If the space (which the author fancies) void of all bodies, is not altogether empty; what is it then full of? Is it full of extended spirits perhaps, or immaterial substances, capable of extending and contracting themselves; which move therein, and penetrate each other without any inconveniency, as the shadows of two bodies penetrate one another upon the surface of a wall? Methinks I see the revival of the odd imaginations of Dr. Henry More (otherwise a learned and well-meaning man,) and of some others who fancied that those spirits can make themselves impenetrable whenever they please. Nay, some have fancied, that man, in the state of innocency, had also the gift of penetration; and that he became solid, opaque, and impenetrable by his fall. Is it not overthrowing our notions of things, to make God have parts, to make spirits have extension? The principle of the want of a sufficient reason does alone drive away all these

spectres of imagination. Men easily run into fictions, for want of making a right use of that great principle.

．　　．　　．　　．　　．

To Paragraph 13

52. In order to prove that space, without bodies, is an absolute reality; the author objected, that a finite material universe might move forward in space. I answered, it does not appear reasonable that the material universe should be finite; and, though we should suppose it to be finite; yet 'tis unreasonable it should have motion any otherwise, than as its parts change their situation among themselves; because such a motion would produce no change that could be observed, and would be without design. 'Tis another thing, when its parts change their situation among themselves; for then there is a motion in space; but it consists in the order of relations which are changed. The author replies now, that the reality of motion does not depend upon being observed; and that a ship may go forward, and yet a man, who is in the ship, may not perceive it. I answer, motion does not indeed depend upon being observed; but it does depend upon being possible to be observed. There is no motion, when there is no change that can be observed. And when there is no change that can be observed, there is no change at all. The contrary opinion is grounded upon the supposition of a real absolute space, which I have demonstratively confuted by the principle of the want of a sufficient reason of things.

53. I find nothing in the Eighth Definition of the *Mathematical Principles of Nature,* nor in the Scholium belonging to it, that proves, or can prove, the reality of space in itself. However, I grant there is a difference between an absolute true motion of a body, and a mere relative change of its situation with respect to another body. For when the immediate cause of the change is in the body, that body is truly in motion; and then the situation of other bodies, with respect to it, will be changed consequently, though the cause of that change be not in them. 'Tis true that, exactly speaking, there is not any one body, that is perfectly and entirely at rest; but we frame an abstract notion of rest, by considering the thing mathematically. Thus have I left nothing unanswered, of what has been alleged for the absolute reality of

space. And I have demonstrated the falsehood of that reality, by a fundamental principle, one of the most certain both in reason and experience; against which, no exception or instance can be alleged. Upon the whole,[18] one may judge from what has been said that I ought not to admit a moveable universe; nor any place out of the material universe.

. . . . .

61. I shall not enlarge here upon my opinion explained elsewhere, that there are no created substances wholly destitute of matter. For I hold with the ancients, and according to reason, that angels or intelligences, and souls separated from a gross body, have always subtile bodies, though they themselves be incorporeal. The vulgar philosophy easily admits all sorts of fictions: mine is more strict.

62. I don't say that matter and space are the same thing. I only say, there is no space, where there is no matter; and that space in itself is not an absolute reality. Space and matter differ, as time and motion. However, these things, though different, are inseparable.

63. But yet it does not at all follow, that matter is eternal and necessary; unless we suppose space to be eternal and necessary: a supposition ill grounded in all respects.

To Paragraph 19

71. The author repeats here what has been already confuted above, Numb. 21; that matter cannot be created, without God's choosing among indiscernibles. He would be in the right, if matter consisted of atoms, similar particles, or other the like fictions of superficial philosophy. But that great principle, which proves there is no choice among indiscernibles, destroys also these ill-contrived fictions.

Space, Time, and Measurement

ROGER JOSEPH BOSCOVICH, from
A Theory of Natural Philosophy

In the mid-eighteenth century, the Serbo-Croatian physicist Roger
J. Boscovich (1711–1787) articulated a theory of motion which
was clearly Newtonian in inspiration. He developed the view that
an Absolute Space and Time exist, which are essentially similar
to those entities proposed by Newton. But he also gave a remarkable analysis of the concept of measurement which revealed the
subtle relations between points of Absolute Space and the bodies
which occupy them.

According to Boscovich, all bodies are composed of material
points which have mass or inertia, but have no dimensions. In this
latter respect they resemble the points of geometry. Boscovich's
theory of motion required that material points attract or repel
each other, depending upon the distance between them. These
points of matter should never be confused with the points of
space. Each point of matter at a given instant of time can occupy
only one point of space, and the distance between the point
bodies is determined by the distance between the points of space
which they occupy. Therefore, we have a very sharp distinction
between bodies, on the one hand, and the parts of Absolute Space,
on the other.

The distinction between spatial relations of points and the
spatial relations of bodies created a problem which every Newtonian had to resolve. If the distance between two bodies is the
distance between the points of space which they occupy, then how
do we gauge the distance between two points of Absolute Space?
Newtonians agreed that the points of space could not be observed
directly. How do we judge and compare distances between points
of Absolute Space? One traditional answer suggests moving a rigid
rod of constant length to see if it fits exactly between each of the

Reprinted by permission of the publisher from *A Theory of Natural Philosophy Put Forward and Explained*, by Roger Joseph Boscovich, S.J. The
Open Court Publishing Company, La Salle, Illinois, 1922.

two pairs of points. But how do we know that the rod keeps the same length throughout the motion? Boscovich even argued that, according to his theory of matter, it is very unlikely that the rod would behave so conveniently. Although this rod or standard is inadequate, according to Boscovich, he believed nevertheless that some assumption about the behavior of bodies was necessary. Otherwise there could be no comparison of the distances or the magnitudes of Absolute Space.

This argument is, as far as we know, the only analysis of measurement given by an adherent of Newtonian theory. By it, Boscovich showed how subtly the science of spatial and temporal magnitudes (of Absolute Space and Time) depends upon certain common but indispensable assumptions about the behavior of sensible bodies. It makes it impossible to confirm any theory of Absolute Space without having a theory of the behavior of moving bodies.

Of Space and Time

1. I do not admit perfectly continuous extension of matter; I consider it to be made up of perfectly indivisible points, which are non-extended, set apart from one another by a certain interval, & connected together by certain forces that are at one time attractive & at another time repulsive, depending on their mutual distances. Here it is to be seen, with this theory, what is my idea of space, & of time, how each of them may be said to be continuous, infinitely divisible, eternal, immense, immovable, necessary, although neither of them, as I have shown in a note, have a real nature of their own that is possessed of these properties.

2. First of all it seems clear to me that not only those who admit absolute space, which is of its own real nature continuous, eternal & immense, but also those who, following Leibniz & Descartes, consider space itself to be the relative arrangement which exists amongst things that exist, over and above these existent things; it seems to me, I say, that all must admit some mode of existence that is real & not purely imaginary; through which they are where they are, & this mode exists when they are there, & perishes when they cease to be where they were. For, such a space being admitted in the first theory, if the fact that there is some thing in that part of space depends on the thing & space alone; then, as often as the thing existed, & space, we should have the fact

that that thing was situated in that part of space. Again, if, in the second theory, the arrangement, which constitutes position, depended only on the things themselves that have that arrangement; then, as often as these things should exist, they would exist in the same arrangement, & could never change their position. What I have said with regard to space applies equally to time.

3. Therefore it needs must be admitted that there is some real mode of existence, due to which a thing is where it is, & exists then, when it does exist. Whether this mode is called the thing, or the mode of the thing, or something or nothing, it is bound to be beyond our imagination; & the thing may change this kind of mode, having one mode at one time & another at another time.

4. Hence, for each of the points of matter (to consider these, from which all I say can be easily transferred to immaterial things), I admit two real kinds of modes of existence, of which some pertain to space & others to time; & these will be called local & temporal modes respectively. Any point has a real mode of existence, through which it is where it is; & another, due to which it exists at the time when it does exist. These real modes of existence are to me real time & space; the possibility of these modes, hazily apprehended by us, is, to my mind, empty space & again empty time, so to speak; in other words, imaginary space & imaginary time.

5. These several real modes are produced & perish, and are in my opinion quite indivisible, non-extended, immovable & unvarying in their order. They, as well as the positions & times of them, & of the points to which they belong, are real. They afford the foundation of a real relation of distance, which is either a local relation between two points, or a temporal relation between two events. Nor is the fact that those two points of matter have that determined distance anything essentially different from the fact that they have those determined modes of existence, which necessarily alter when they change the distance. Those modes which are descriptive of position I call real points of position; & those that are descriptive of time I call instants; & they are without part, & the former lack any kind of extension, while the latter lack duration; both are indivisible.

6. Further, a point of matter that is perfectly indivisible & non-extended cannot be contiguous to any other point of matter; if they have no distance from one another, they coincide com-

pletely; if they do not coincide completely, they have some distance between them. For, since they have no kind of parts, they cannot coincide partly only; that is, they cannot touch one another on one side, & on the other side be separated. It is but a prejudice acquired from infancy, & born of ideas obtained through the senses, which have not been considered with proper care; & these ideas picture masses to us as always being composed of parts at a distance from one another. It is owing to this prejudice that we seem to ourselves to be able to bring even indivisible and non-extended points so close to other points that they touch them & constitute a sort of lengthy series. We imagine a series of little spheres, in fact; & we do not put out of mind that extension, & the parts, which we verbally exclude.

7. Again, where two points of matter are at a distance from one another, another point of matter can always be placed in the same straight line with them, on the far side of either, at an equal distance; & another beyond that, & so on without end, as is evident. Also another point can be placed halfway between the two points, so as to touch neither of them; for, if it touched either of them it would touch them both, & thus would coincide with both; hence the two points would coincide with one another & could not be separate points, which is contrary to the hypothesis. Therefore that interval can be divided into two parts; & therefore, by the same argument, those two can be divided into four others, & so on without any end. Hence it follows that, however great the interval between two points, we could always obtain another that is greater; &, however small the interval might be, we could always obtain another that is smaller; &, in either case, without any limit or end.

8. Hence beyond & between two real points of position of any sort there are other real points of position possible; & these recede from them & approach them respectively, without any determinate limit. There will be a real divisibility to an infinite extent of the interval between two points, or, if I may call it so, an endless 'insertibility' of real points. However often such real points of position are interpolated; by real points of matter being interposed, their number will always be finite, the number of intervals intercepted on the first interval, & at the same time constituting that interval, will be finite; but the number of possible parts of this sort will be endless. The magnitude of each

of the former will be definite & finite; the magnitude of the latter will be diminished without any limit whatever; & there will be no gap that cannot be diminished by adding fresh points in between; although it cannot be completely removed either by division or by interposition of points.

9. In this way, so long as we conceive as possibles these points of position, we have infinity of space, & continuity, together with infinite divisibility. With existing things there is always a definite limit, a definite number of points, a definite number of intervals; with possibles, there is none that is finite. The abstract concept of possibles, excluding as it does a limit due to a possible increase of the interval, a decrease or a gap, gives us the infinity of an imaginary line, & continuity; such a line has not actually any existing parts, but only possible ones. Also, since this possibility is eternal, in that it was true from eternity & of necessity that such points might exist in conjunction with such modes, space of this kind, imaginary, continuous & infinite, was also at the same time eternal & necessary; but it is not anything that exists, but something that is merely capable of existing, & an indefinite concept of our minds. Moreover, immobility of this space will come from immobility of the several points of position.

10. Everything, that has so far been said with regard to points of position, can quite easily in the same way be applied to instants of time; & indeed there is a very great analogy of a sort between the two. For, a point from a given point, or an instant from a given instant, has a definite distance, unless they coincide; & another distance can be found either greater or less than the first, without any limit whatever. In any interval of imaginary space or time, there is a first point or instant, & a last; but there is no second, or last but one. For, if any particular one is supposed to be the second, then, since it does not coincide with the first, it must be at some distance from it; & in the interval between, other possible points or instants intervene. Again, a point is not a part of a continuous line, or an instant a part of a continuous time; but a limit & a boundary. A continuous line, or a continuous time is understood to be generated, not by repetition of points or instants, but by a continuous progressive motion, in which some intervals are parts of other intervals; the points themselves, or the instants, which are continually progressing, are not parts of the intervals. There is but one difference, namely,

that this progressive motion can be accomplished in space, not only in a single direction along a line, but in infinite directions over a plane which is conceived from the continuous motion of the line already conceived in the direction of its breadth; & further, in infinite directions throughout a solid, which is conceived from the continuous motion of the plane already conceived. Whereas, in time there will be had but one progressive motion, that of duration; & therefore this will be analogous to a single line. Thus, while for imaginary space there is extension in three dimensions, length, breadth & depth, there is only one for time, namely length or duration only. Nevertheless, in the threefold class of space, & in the onefold class of time, the point & the instant will be respectively the element, from which, by its progression, motion, space & time will be understood to be generated.

11. Now here there is one thing that must be carefully noted. Not only when two points of matter exist, & have a distance from one another, do two modes exist which give the foundation of the relation of this distance; & there are two different real points of position, the possibility of which, as conceived by us, will yield two points of imaginary space; & thus, to the infinite number of possible points of matter there will correspond an infinite number of possible modes of existence. But also to any one point of matter there will correspond the infinite possible modes of existing, which are all the possible positions of that point. All of these taken together are sufficient for the possession of the whole of imaginary space; & any point of matter has its own imaginary space, immovable, infinite & continuous; nevertheless, all these spaces, belonging to all points coincide with one another, & are considered to be one & the same. For if we take one real point of position belonging to one point of matter, & associate it with all the real points of position belonging to another point of matter, there is one among the latter, which, if it coexist with the former, will induce a relation of no-distance, which we call compenetration. From this it is clear that, for points which exist, no-distance is not nothing, but a relation induced by some two modes of existence. Any of the others would induce, with that same former point of position, another relation of some determinate distance & position, as we say. Further, those points of position, which induce a relation of no-distance, we consider to

be the same; & we consider any of the infinite number of such points belonging to the infinite number of points of matter to be the same; & mean them when we speak of the 'same position.' Moreover this is evidently bound to be true for any pair of points. If now a third point is situated anywhere, it will have some distance & position with respect to the first. If the first is removed, the second can be so situated that it has the same distance & position with respect to the third as the first had. Hence the mode, in which it exists, will be taken to be the same in this case as the mode in which the first point was existing; & if these two modes were existing together, they would induce a relation of no-distance between the first point & the second. All that has been said above with regard to points of space applies equally well to instants of time.

12. Now, whether they can coexist is a question that pertains to the relation between points of position & instants of time, whether we consider a single point of matter or several of them. In the first place, several instants of time belonging to the same point of matter cannot coexist; but they must necessarily come one after the other; & similarly, two points of position belonging to the same point of matter cannot be conjoined, but must lie one outside the other; & this comes from the nature of points of this kind, & is essential to them, to use a common phrase.

·   ·   ·   ·   ·

Of Space & Time, as we know them

18. We have spoken, in the preceding Supplement, of Space & Time, as they are in themselves; it remains for us to say a few words on matters that pertain to them, in so far as they come within our knowledge. We can in no direct way obtain a knowledge through the senses of those real modes of existence, nor can we discern one of them from another. We do indeed perceive, by a difference of ideas excited in the mind by means of the senses, a determinate relation of distance & position, such as arises from any two local modes of existence; but the same idea may be produced by innumerable pairs of modes or real points of position; these induce the relations of equal distances & like positions, both amongst themselves & with regard to our organs, & to the rest of the circumjacent bodies. For, two points of matter,

which anywhere have a given distance & position induced by some two modes of existence, may somewhere else on account of two other modes of existence have a relation of equal distance & like position, for instance if the distances exist parallel to one another. If those points, we, & all the circumjacent bodies change their real positions, & yet do so in such a manner that all the distances remain equal & parallel to what they were at the start, we shall get exactly the same ideas. Nay, we shall get the same ideas, if, while the magnitudes of the distances remain the same, all their directions are turned through any the same angle, & thus make the same angles with one another as before. Even if all these distances were diminished, while the angles remained constant, & the ratio of the distances to one another also remained constant, but the forces did not change owing to that change of distance; then if the scale of forces is correctly altered, that is to say, that curved line, whose ordinates express the forces; then there would be no change in our ideas.

19. Hence it follows that, if the whole Universe within our sight were moved by a parallel motion in any direction, & at the same time rotated through any angle, we could never be aware of the motion or the rotation. Similarly, if the whole region containing the room in which we are, the plains & the hills, were simultaneously turned round by some approximately common motion of the Earth, we should not be aware of such a motion; for practically the same ideas would be excited in the mind. Moreover, it might be the case that the whole Universe within our sight should daily contract or expand, while the scale of forces contracted or expanded in the same ratio; if such a thing did happen, there would be no change of ideas in our mind, & so we should have no feeling that such a change was taking place.

20. When either objects external to us, or our organs change their modes of existence in such a way that that first equality or similitude does not remain constant, then indeed the ideas are altered, & there is a feeling of change; but the ideas are the same exactly, whether the external objects suffer the change, or our organs, or both of them unequally. In every case our ideas refer to the difference between the new state & the old, & not to the absolute change, which does not come within the scope of our senses. Thus, whether the stars move round the Earth, or the Earth & ourselves move in the opposite direction round them,

the ideas are the same, & there is the same sensation. We can never perceive absolute changes; we can only perceive the difference from the former configuration that has arisen. Further, when there is nothing at hand to warn us as to the change of our organs, then indeed we shall count ourselves to have been unmoved, owing to a general prejudice for counting as nothing those things that are nothing in our mind; for we cannot know of this change, & we attribute the whole of the change to objects situated outside of ourselves. In such manner any one would be mistaken in thinking, when on board ship, that he himself was motionless, while the shore, the hills & even the sea were in motion.

21. Again, it is to be observed first of all that from this principle of the unchangeability of those things, of which we cannot perceive the change through our senses, there comes forth the method that we use for comparing the magnitudes of intervals with one another; here, that, which is taken as a measure, is assumed to be unchangeable. Also we make use of the axiom, *things that are equal to the same thing are equal to one another;* & from this is deduced another one pertaining to the same thing, namely, *things that are equal multiples, or submultiples, of each, are also equal to one another;* & also this, *things that coincide are equal.* We take a wooden or iron ten-foot rod; & if we find that this is congruent with one given interval when applied to it either once or a hundred times, & also congruent to another interval when applied to it either once or a hundred times, then we say that these intervals are equal. Further, we consider the wooden or iron ten-foot rod to be the same standard of comparison after translation. Now, if it consisted of perfectly continuous & solid matter, we might hold it to be exactly the same standard of comparison; but in my theory of points at a distance from one another, all the points of the ten-foot rod, while they are being transferred, really change the distance continually. For the distance is constituted by those real modes of existence, & these are continually changing. But if they are changed in such a manner that the modes which follow establish real relations of equal distances, the standard of comparison will not be identically the same; & yet it will still be an equal one, & the equality of the measured intervals will be correctly determined. We can no more transfer the length of the ten-foot rod, constituted in its first

position by the first real modes, to the place of the length constituted in its second position by the second real modes, than we
are able to do so for intervals themselves, which we compare by
measurement. But, because we perceive none of this change
during the translation, such as may demonstrate to us a relation
of length, therefore we take that length to be the same. But
really in this translation it will always suffer some slight change.
It might happen that it underwent even some very great change,
common to it & our senses, so that we should not perceive the
change; & that, when restored to its former position, it would
return to a state equal & similar to that which it had at first.
However, there always is some slight change, owing to the fact
that the forces which connect the points of matter, will be
changed to some slight extent, if its position is altered with
respect to all the rest of the Universe. Indeed, the same is the
case in the ordinary theory. For no body is quite without little
spaces interspersed within it, altogether incapable of being compressed or dilated; & this dilatation & compression undoubtedly
occurs in every case of translation, at least to a slight extent. We,
however, consider the measure to be the same so long as we do
not perceive any alteration, as I have already remarked.

22. The consequence of all this is that we are quite unable to
obtain a direct knowledge of absolute distances; & we cannot
compare them with one another by a common standard. We
have to estimate magnitudes by the ideas through which we
recognize them; & to take as common standards those measures
which ordinary people think suffer no change. But philosophers
should recognize that there is a change; but, since they know
of no case in which the equality is destroyed by a perceptible
change, they consider that the change is made equally.

23. Further, although the distance is really changed when, as
in the case of the translation of the ten-foot rod, the position of
the points of matter is altered, those real modes which constitute
the distance being altered; nevertheless if the change takes place
in such a way that the second distance is exactly equal to the
first, we shall call it the same, & say that it is altered in no way,
so that the equal distances between the same ends will be said
to be the same distance & the magnitude will be said to be the
same; & this is defined by means of these equal distances, just as
also two parallel directions will be also included under the name

of the same direction. In what follows we shall say that the distance is not changed, or the direction, unless the magnitude of the distance, or the parallelism, is altered.

24. What has been said with regard to the measurement of space, without difficulty can be applied to time; in this also we have no definite & constant measurement. We obtain all that is possible from motion; but we cannot get a motion that is perfectly uniform. We have remarked on many things that belong to this subject, & bear upon the nature & succession of these ideas, in our notes. I will but add here, that, in the measurement of time, not even ordinary people think that the same standard measure of time can be translated from one time to another time. They see that it is another, consider that it is an equal, on account of some assumed uniform motion. Just as with the measurement of time, so in my theory with the measurement of space it is impossible to transfer a fixed length from its place to some other, just as it is impossible to transfer a fixed interval of time, so that it can be used for the purpose of comparing two of them by means of a third. In both cases, a second length, or a second duration is substituted, which is supposed to be equal to the first; that is to say, fresh real positions of the points of the same ten-foot rod which constitute a new distance, such as a new circuit made by the same rod, or a fresh temporal distance between two beginnings & two ends. In my Theory, there is in each case exactly the same analogy between space & time. Ordinary people think that it is only for measurement of space that the standard of measurement is the same; almost all other philosophers except myself hold that it can at least be considered to be the same from the idea that the measure is perfectly solid & continuous, but that in time there is only equality. But I, for my part, only admit in either case the equality, & never the identity.

On Absolute Space and Time

LEONHARD EULER, from *Reflections on Space and Time*

Leonhard Euler (1707–1783) is justly celebrated as one of the greatest mathematicians and physicists of the eighteenth century. His discoveries in number theory and topology are well known, and he solved many problems in physics which his contemporaries found difficult even to formulate. Euler gave the Newtonian laws of motion a formulation which used the notion of a differential equation—a form in which they are still taught today. He was also able to reformulate and prove the Principle of Least Action, a principle which has had wide application beyond the confines of Newtonian physics. There was no question, therefore, that when Euler spoke of the achievements and requirements of the physics of his day, he knew whereof he spoke.

Euler, according to an insight of Ernst Casirer, was one of the first to reverse a traditional relation between physics and philosophy. Before his work, it was customary to think that no concept was to be used in physics or science generally unless it met certain standards of metaphysics. But in Euler's time, metaphysics spoke with two voices about the concepts of space and time. As we have seen, some philosophers like Newton, Clarke, and Boscovich thought that space (and time) were entities having extensionless parts (that is points and instants, respectively); others, like Leibniz, believed that space was a special ordering of coexistent bodies and that time, too, was a special order of bodies. Euler's reverse consisted in his assertion that the laws of physics, by which he meant Newton's laws of motion, were so well confirmed that no one would be right to deny them. These laws not only state truths about the world, but they also explain why things happen; that is, they give the causes of certain kinds of events. Euler argued that if one followed philosophers like Leibniz and used a relational theory of space, two consequences would follow: (1)

Reprinted by permission from Link M. Lotter: *The Philosophical Significance of Leonhard Euler,* Master's Essay, Faculty of Philosophy, Columbia University, 1929.

substituting Leibniz's concept of space for Newton's transforms
the true laws of Newton into demonstrably false statements, and
(2) we cannot explain why certain kinds of events occur, if we
employ Leibniz's concept of space rather than Newton's. Granted,
then, that we do have true laws and explanations, the conclusion
is, clearly, that both Leibniz and Descartes did not have the
proper understanding of space and time.

Euler's argument seems to show that the Newtonian concepts
are certainly preferable to those of Leibniz, if we believe that we
have true laws and explanations which involve space and time.
In the course of this remarkable defense of Newton's concept of
Absolute Space and Time, Euler suggested several ideas which
were to have great influence through the works of Immanuel
Kant (pp. 126, 133). The major thrust of Euler's study lies in its
suggestion that the ideas and principles of metaphysics ought to
be regulated and determined by the knowledge which physics has
established, and not the other way around.

This essay originally appeared in *The History of the Royal
Berlin Academy of Sciences,* 1748.

I

The principles of mechanics have already been established on
such a sound basis that one would greatly err if he wished to
encourage any doubt as to their validity. Even if one were not
in position to demonstrate them by the use of the general principles of metaphysics, the excellent agreement of all the conclusions which one draws from them by means of the calculus, with
all the movements of bodies both solids and liquids, on the
earth, and likewise with the movements of the heavenly bodies,
would be sufficient to place the truth of the principles of mechanics, beyond doubt. Thus it is an unquestionable fact, that
a body being once at rest will remain continually at rest unless
it be disturbed in its state of rest by some external force. It is
also equally certain, that a body, being once set in motion, will
continually move with the same speed in the same direction provided that it does not meet with obstacles contrary to the preservation of that state.

II

Since these two truths are so certainly verified, it follows with
absolute necessity that they depend upon the nature of bodies:
and since it is metaphysics which is concerned in investigating

the nature and properties of bodies, the knowledge of these truths of mechanics is capable of serving as a guide in these intricate researches (of metaphysics). For one would be right in rejecting in this science (of metaphysics) all the reasons and all the ideas, however well founded they may otherwise be, which lead to conclusions contrary to these truths (of mechanics); and one would be warranted in not admitting any such principles which cannot agree with these same truths. The first ideas which we form for ourselves of things, which are found outside of ourselves, are ordinarily so obscure and so indefinite that it is extremely unsafe to draw from them conclusions of which one can be certain. Thus it is always a great step in advance when one already knows some conclusions from some other source, at which the first principles of metaphysics ought to finally arrive: and it will be by these conclusions, that the principal ideas of metaphysics will be necessarily regulated and determined.

III

Also the metaphysicians, very far from denying these principles, concerning the truth of which mechanics assures us, try to deduce and to prove these from their own ideas. But they upbraid the mathematicians, because they apply these principles inappropriately to ideas of space and time which are taken to be only imaginary and destitute of all reality. It is very possible, that a true principle can be expressed in an inappropriate manner, without losing all of its validity, and that it does not agree with the exact ideas which one ought to have of things; but then the metaphysician should be obliged to remedy this defect, and to substitute in the expression of these principles real ideas instead of imaginary ones.

IV

Such then would be the case concerning these principles of mechanics, which are involved in the ideas of space and time, which according to the metaphysicians do not have any reality: thus it becomes necessary to see, whether or not it is possible to separate from them these imaginary ideas and to substitute in their places the real ideas, from which we have formed these imaginary ideas by way of abstraction: of a kind nevertheless that would not in the least alter the significance and force of these principles. For there is not the least doubt, that bodies of themselves behave in accordance with these principles, and not in any sense are regulated by things which exist only in our imagination: it is moreover certain, that there are those very real things, to which these laws refer, that bodies persevere in the maintenance of their state.

V

Thus it is certain, that if it is not possible to conceive the two principles adduced from mechanics without their being involved in the ideas of space and time, this is a sure indication that these ideas are not purely imaginary as the metaphysicians suppose. One is entitled moreover to conclude from it, that both absolute space, and time, such as are represented by mathematicians, are real things, which also exist outside of our imagination: since it would be absurd to hold that pure fictions of the mind can serve as the basis for real principles of mechanics.

VI

To enter into this investigation, I shall begin with the first principle which concerns the state of rest of bodies. In mechanics, space and place are considered as real things, and by this principle one holds that a body, which exists in any place without motion, will continually remain in it unless it be expelled from it by some external force: in this instance then this body will remain always in the same place with reference to absolute space. I am most willing to let the ideas of space and of place be only imaginary notions; but that I may be shown the realities, by which bodies are ordered in obeying this law; and in place of which the mathematicians are satisfied in providing for themselves imaginary ideas of space and of place.

VII

One will say to me at the very outset, that place is nothing else than the relation of one body in reference to others which surround it. Let us substitute then this idea for that of place, and one will be obliged to say, that by virtue of this principle a body existing at one time in one certain relation to other bodies, which surround it, will persistently strive to continue always in this same relation. That is to say, one ought to hold that a body A being surrounded by bodies B, C, D, E, etc. will endeavor to keep itself always in this same neighborhood. And in objecting to what the mathematician says, that a body at rest, remains in the same place, with reference to absolute space: the metaphysician will say, that this body preserves itself in the same relation of reference to other bodies which surround it.

VIII

Let us see whether these two ways of expression are equivalent, and whether one can always, without falling into error, substitute the metaphysical expression for the mathematical one, concerning the truth of which we are already convinced. Suppose then in order to put these two expressions in accord, that both body A and its neighbors B, C, D, etc. are at rest; and in this case the body A in keeping itself in the same neighborhood of bodies B, C, D, E, etc., according to the metaphysical rule, will continue also in the same place according to the mathematical rule; and in this instance one will not be deceived in substituting the latter in place of the former.

IX

Suppose, in order to better arrange our ideas, that the body A is in a pool of still water, and while it remains in the same place, it will also remain in the same neighborhood of the particles of water, which surround it, and this body will be ordered equally by the rule of mathematics and by that of metaphysics. But suppose now, that the water begins to flow, and according to the rule of mathematics the body will still remain in the same place unless it is carried away gradually by the force of the water. Now according to the metaphysical rule this body ought at once to follow completely the movement of the water in order to maintain itself in the neighborhood of the same particles of water which surrounded it previously. In this case then the rule derived from metaphysics no longer conforms to the truth.

X

If we consult experience in this matter, we will learn that a body, having been at rest in a pool of still water, will be set in motion as soon as the water begins to flow, which appears to favor the rule according to metaphysics. But mechanics makes us see very clearly that the body does not follow the current of water, but instead that it is struck by the particles of water; and that it is consequently, an external force which sets the body in motion. Then without this force the body would likewise rightly remain at rest, in the current of water as in the still water, and consequently the body in the maintenance of its own state of rest is not in the least regulated by the bodies which immediately surround it. From this it follows that what is called place in mechanics does not allow the explanation of metaphysics by which one admits that place is nothing else than the relation of a body in reference to other bodies which surround it.

XI

To this property of bodies, by virtue of which they try to maintain themselves in their state, as well of rest as of motion, one gives the name inertia. Then this inertia, as we come to see, is not determined at all by the surrounding bodies; but it is very certain that it is determined by the idea of place, which the mathematicians consider as real, and the metaphysicians as imaginary. Thus since we are not permitted to substitute for this idea of place, the relation of a body to surrounding bodies, it does not stand that the remote bodies are those by reference to which one is able to judge of the general principle of inertia. But I doubt strongly whether the metaphysicians would risk maintaining that bodies by virtue of their inertia are disposed to preserve the same relation in reference to other bodies, which are removed to some distance from them: for it would be easy to illustrate the falsity of such an explanation by similar reflections, which I have proceeded to make about the bodies of the immediate neighborhood.

XII

If they should say that it was in reference to the fixed stars which was necessary for explaining the principle of inertia: it would be very difficult to refute this, since the fixed stars, being likewise at rest, are so distant from us, that the bodies which exist at rest in reference to absolute space, as one considers it in mathematics, would be also at rest in reference to the fixed stars. But outside of that, which would be a proposition very strange and contrary to a large portion of the other doctrines of metaphysics, to say, that the fixed stars govern bodies in their inertia, this rule would be found equally false, if we are not permitted to apply it to bodies which are near to some fixed star. Having observed these things, no more of the real ideas, which one could substitute in place of these supposedly imaginary ideas of space and of place, in the explanation of inertia.

XIII

We see then that the idea of place, such as mathematicians conceive it, cannot be explained by any relation to other bodies, either near or remote, and consequently the metaphysical notions, which one believes to correspond with the mathematical idea of place, are not appropriate to be introduced into the explanation of the mechanical principle in question. This principle relating to the preservation of the state of bodies is determined by the place, such as one conceives it in mathematics, and not at all by reference to other bodies. Now one cannot say that this principle of mechanics is founded on anything, which exists only in the imagination: and from this it is necessary to conclude absolutely that the mathematical idea of place is not imaginary, but that there is such a real thing in the world, which corresponds to that idea. There is then in the world, outside of the bodies which constitute it, some reality, which we represent to ourselves by the idea of place.

XIV

So the metaphysicians are wrong, when they want to banish from the world space and place, in maintaining that these are only abstract and imaginary ideas. Consequently the proofs that they produce for holding their opinion, however well grounded they may be, are in effect poor reasons and it is necessary that anything fallacious be concealed. It is true that the senses are not capable of furnishing us the ideas of space and of place, and that it is only through reflection, that we form these ideas. From this they conclude that these are only abstract ideas, similar to those of genus and of species, which exist only in our understanding, and to which no real object corresponds. But it appears to me that this conclusion is hasty: for from a little reflection on one's self, one will easily see that the manner, in which one arrives at the idea of space and of place, is very different from that, in which we form the ideas of genus and of species. And one is decidedly mistaken, if he wishes to hold, that those things do not exist, of which we have no ideas other than through reflection.

XV

I am of the opinion that all the things which exist are completely determined; and if we take away from one such object, one or more determinations, that from it is born a generic idea, to which there does not correspond any existing object. This is how we form the idea of extension in general, in deducting from the ideas of bodies all the determinations, except extension. But the idea of the place which a body occupies, is not formed in subtracting certain determinations from bodies; it results in removing the body entirely: in such a manner that the place has not been a determination of the body, after having removed the body, entirely with all of its properties. For it is necessary to observe that the place which a body occupies is very different from its extension, because extension belongs to body and changes with it by movement from one place to another; but that place and space are not susceptible to such movement.

XVI

I do not wish to enter into the discussion of the objections which are made against the reality of space and of place; for since it has been shown, that this reality can no longer be questioned, it follows necessarily, that all these objections ought to have little weight; although we might also be unable to make reply. If one thinks it absurd, that all the different places, or parts of space are similar hollows, this would be contrary to the principle of indiscernibility; I do not know whether this principle is as general as it is thought to be; perhaps it is not applicable to both bodies and spirits; as to generality, one can well be content: but as space and place are things so essentially different from spirits and bodies, one could not judge of it by the same principles.

XVII

The reality of space will be found again established through the other principle of mechanics, which involves the preservation of uniform motion in the same direction. For if space and place are not the reference of co-existing bodies, what is meant by the same direction? One would be very much confused in expressing an idea of the mere relation of mutually co-existing bodies, without making use of that of immovable space. For in some manner, while bodies move and change the positions of their situation, that does not prevent one from preserving a distinctly clear idea of one fixed direction which bodies tend to follow in their movement, in spite of all the alterations which other bodies experience. From this it is evident, that the identity of direction, which is a distinctly essential factor in the general principles of motion, cannot be absolutely explained by the relation, or the order of co-existing bodies. Thus again there must necessarily be some other real existence, outside the bodies, to which the idea of one same direction corresponds; and there is no doubt that this would be the space, the reality of which we have established.

XVIII

The ideas of space and of time have nearly always had the same fate, in this way that those who have denied the reality of one, have also denied that of the other, and vice versa. One would not then be surprised, that in establishing the reality of space, we recognize time also, as something real, which exists not only in our mind, but which actually flows in serving as a measure for the duration of things. We have a very definite idea of time, and I agree that we form it from the successions of changes which we observe: in this sense I stand agreed, that the idea of time does not exist outside of our imagination. But there is room to ask, whether the idea of time, and time itself, are not in themselves different things? And it seems to me that the metaphysicians, in destroying the reality of time, have confounded time itself with the idea which we have of it.

XIX

The principle of the motion of bodies, by virtue of which, a body set in motion is obliged to continue with the same velocity in the same direction, this principle, I say, furnishes us with new proofs, not only for the reality of space, but also for that of time. For, since the uniform motion traverses equal spaces in equal times I ask first of all, what it is which is equal in regard to spaces, according to the opinion of those who deny the reality of space? I am very much in doubt whether the metaphysicians would risk saying that the equality of spaces ought to be judged by the equality of the number of monads which fill them: for they ought to hold that the monads would be equally distributed throughout all bodies. But even if they wished to hold to this explanation, it would be at once reversed the moment that the motion of bodies, with reference to which one wishes to determine the equality of spaces, is taken into consideration. For we conceive, and the principle of motion shows us, that when a body traverses equal spaces, the equality of spaces does not depend in any way on other bodies which surround it, and that remains the same for any alterations which other bodies undergo.

XX

It is the same with the equality of times; for if time is nothing else, as is required in metaphysics, than the order of successions, in what way is the equality of times rendered intelligible? It is taken for granted that every being in the world is subject to continuous changes, and that it is the succession of these changes, which gives rise to time. According to this explanation, two times ought to be equal, during which the same number of successions occur. But if one considers a body, which traverses equal spaces in equal times, by what changes or by what body is it necessary to judge the equality of these two times? Or it may be granted that all bodies are subjected to the same changes equally often, in such a manner that it would be all one whatever body one wishes to choose for the purpose of measuring the equality of times by the number of changes which take place. But I am sure, that if one will weigh this explanation a little, he will find all kinds of inconveniences, so that he will easily be persuaded to abandon it.

XXI

The question here is not concerning our estimate of the equality of times, which doubtless depends on the state of our minds; the question concerns the equality of times, during which a body moving with uniform motion traverses equal spaces. Since this equality cannot be explained by the order of successions, so just as little can the equality of spaces be explained by the order of co-existent ones, and whether it enters essentially into the principle of motion; it cannot be said that the bodies, in pursuing their motion, are regulated by something that exists only in our imagination. One would then be obliged to admit, as he has been in reference to space, that time is something, which exists outside of our mind, or that time is something real, as properly as space. I address myself here to those metaphysicians who still recognize some reality in bodies and in motion; as for those who deny this reality altogether, and who do not admit those phenomena, since they regard both motion itself, and the laws of motion as idle fancies, I do not flatter myself that these reflections would make the least impression on their mind.

On the Reality of Space

IMMANUEL KANT, from *Regions in Space*

Immanuel Kant (1724–1804) exerted an enormous influence upon both philosophers' and physicists' ways of understanding their subjects. He was greatly influenced by many of Euler's arguments for the view that the senses cannot *furnish* us with ideas of space and time. However his purpose was quite different from Euler's. Euler, it will be recalled (Euler, p. 115), wanted to show that the Newtonian concepts of Absolute Space and Time were preferable to those of Leibniz and Descartes. Rather than support one view or the other, Kant wished to defend a third position, which he believed was preferable to both.

According to Kant, Euler tried to establish the existence of space (and time) by considering certain truths of mechanics such as the Law of Inertia. Kant, however, believed that he could establish the existence of space and time by considering our ordinary experience with bodies; that is, by a consideration of the truths of geometry.

Thus, Kant asks us to consider the difference between a left hand and a right hand, a special case of two bodies which are called incongruous counterparts. A left-handed glove would fit the left hand but not the right hand. How do we explain the difference between the two bodies? The difference cannot be explained by appealing to the way the parts of these bodies are organized, since the internal organization for each is identical. For example, the thumb is related to the four fingers exactly the same way for both left and right hands. One might try to explain the difference by showing that each of the hands is differently related to a third body—perhaps a mirror. But Kant's answer would be plain enough: the two systems, left hand plus mirror and right hand plus mirror, are again incongruous counterparts. To indicate the difference between them would require a fourth body. And the regression could be repeated indefinitely.

Reprinted by permission of the publisher from *Kant's Inaugural and Early Writings on Space*. Tr. by John Handyside. The Open Court Publishing Company, La Salle, Illinois, 1929.

Kant drew the conclusion that the distinction between incongruous counterparts can be explained only by reference to an entity which is not a body; that is, to space. This argument, much like Euler's argument about distinguishing between the same and different directions of motion is directed against both Leibniz and Descartes. But Kant did not thereby accept the Newtonian concepts. He argued that space and time exist. Further, space cannot be an object, a body; it cannot be a property of bodies, nor can it be a relation between bodies. Kant thought Leibniz was surely wrong in believing the latter view. Space is not some immense nonbody in which all bodies are contained and through which they move without resistance. Space, according to Kant, is a schema for the coordination of all our outer sensa or perceptions. (Time is a schema for the coordination by all our inner perceptions.) Kant tells us that this rule by which our external or outer perceptions are organized is such that the rules of geometry are satisfied. "Nature," he tells us, "can be revealed to our senses only in accordance with the rules of geometry." He believed that this explains why geometry is not simply an empirical body of truths, but consists of truths which are necessarily true, for which there is no alternative. Kant also seems to have thought that the reason why one schema coordinates our perceptions rather than another is traceable to the fact that we are rational creatures.

None of these claims is supposed to be self-evident. Kant devoted many detailed arguments to make these theses convincing. But it is clear enough that this view of space and time is not Newtonian, even though it may sometimes appear that way. For example, Euler (following Newton) maintained that space exists even if there is no matter. That is, as Euler described it, there can be empty space, or space independent of matter. Kant also maintained that space is independent, even prior to matter. By this Kant meant that we, as rational creatures, have the disposition to organize our perceptions of bodies in a very special manner. We have this disposition even when we do not perceive bodies— and we could not organize our perceptions that way unless we already had such a disposition.

Kant's concepts of space and time were designed, in part, to find the true ground between two competing views, Newton's and Leibniz's. His solution is very unlike either, and we leave it to the reader to ponder how successful Kant was in constructing an alternative science to Newton's, using Kantian space and time among its basic concepts. Aside from this issue, however, Kant forced physicists and philosophers to reexamine the concepts

and principles of their explanatory schemes to see what was
necessary and what was empirical; what was merely a matter of
fact, but which could have been otherwise, and what was in principle without any alternative; to understand where our knowledge
was still subject to change, and where we had reached bedrock.

The selections are from two essays. The first, "On the First
Ground of the Distinction of Regions of Space," was published
in 1768. The second, "The Dissertation," appeared in 1770.

. . . my aim in this treatise is to investigate whether there is not
to be found in the intuitive judgments of extension, such as are
contained in geometry, an evident proof *that absolute space has
a reality of its own, independent of the existence of all matter,
and indeed as the first ground of the possibility of the composite-
ness of matter.*
Everybody knows how futile have been the endeavours of the
philosophers, by means of the most abstract propositions of metaphysics, to settle this point once for all; and I know of no attempt, save one, to carry this out *a posteriori* as it were, that is,
by means of other undeniable propositions which, though themselves lying outside the realm of metaphysics, can afford through
their application in the concrete a criterion of their correctness.
The one attempt, to which I have referred, was made by the
celebrated Euler, the elder, in 1748, as recorded in the *History
of the Royal Berlin Academy of Sciences*[1] for that year. But so
far from fully achieving his purpose, he only brings to view the
difficulty of assigning to the most general laws of motion a determinate meaning, should we assume no other concept of space
than that obtained by abstraction from the relation of actual
things. The no less notable difficulties which remain in the application of the aforesaid laws, when we endeavour to represent
them in the concrete according to the concept of absolute space,
are left unconsidered. The proof which I here seek should supply,
not to the mechanists (as Herr Euler intended), but to the geometers themselves, a convincing ground for asserting the actuality
of their absolute space, and should do so with the evidence to
which they are accustomed. With this purpose in view, I make
the following preparatory observations.
In physical space, on account of its three demensions, we can

conceive three planes which intersect one another at right angles. Since through the senses we know what is outside us only in so far as it stands in relation to ourselves, it is not surprising that we find in the relation of these intersecting planes to our body the first ground from which to derive the concept of regions in space.[2] The plane to which the length of our body stands perpendicular is called, in reference to us, horizontal; it gives rise to the distinction of the regions we indicate by *above* and *below*. Two other planes, also intersecting at right angles, can stand perpendicular to this horizontal plane, in such manner that the length of the human body is conceived as lying in the line of their intersection. One of these vertical planes divides the body into two outwardly similar parts and supplies the ground for the distinction between *right* and *left;* the other, which is perpendicular to it, makes it possible for us to have the concept of *before* and *behind.* In a written page, for instance, we have first to note the difference between front and back and to distinguish the top from the bottom of the writing; only then can we proceed to determine the position of the characters from right to left or conversely. Here the parts arranged upon the surface have always the same position relatively to one another, and the parts taken as a whole present always the same outlines howsoever we may turn the sheet. But in our representation of the sheet the distinction of regions is so important, and is so closely bound up with the impression which the visible object makes, that the very same writing becomes unrecognisable when seen in such a way that everything which formerly was from left to right is reversed and is viewed from right to left.

Even our judgments about the cosmic regions are subordinated to the concept we have of regions in general, in so far as they are determined in relation to the sides of the body. All other relations that we may recognise, in heaven and on earth, independently of this fundamental conception, are only positions of objects relatively to one another. However well I know the order of the cardinal points, I can determine regions according to that order only in so far as I know towards which hand this order proceeds; and the most complete chart of the heavens, however perfectly I might carry the plan in my mind, would not teach me, from a known region, North say, on which side to look for sunrise, unless, in addition to the positions of the stars in relation to one another, this region were also determined through the position of the plan relatively to my hands. Similarly, our geographical knowledge, and even our commonest knowledge of the position of places, would be of no aid to us if we could not, by reference to the sides of our bodies, assign to regions the things so ordered and the whole system of mutually relative positions.

. . . . .

What, therefore, we desire to show is that the complete ground of determination of the shape of a body rests not merely upon the position of its parts relatively to one another, but further on a relation to the universal space which geometers postulate — a relation, however, which is such that it cannot itself be immediately perceived. What we do perceive are those differences between bodies which depend exclusively upon the ground which this relation affords. If two figures drawn upon a plane are equal and similar, they can be superimposed. But with physical extension and also with lines and surfaces that do not lie in one plane, the case is often quite different. They can be perfectly equal and similar, yet so different in themselves that the boundaries of the one cannot be at the same time the boundaries of the other. A screw which winds round its axis from left to right will not go into a threaded cylinder whose worm goes from right to left, although the thickness of the stem and the number of turns in an equal length correspond. Two spherical triangles can be perfectly equal and similar, and yet not allow of superposition. But the commonest and clearest example is to be found in the limbs of the human body, which are symmetrically disposed about its vertical plane. The right hand is similar and equal to the left, and if we look at one of them alone by itself, at the proportions and positions of its parts relatively to one another and at the magnitude of the whole, a complete description of it must also hold for the other in every respect.

When a body is perfectly equal and similar to another, and yet cannot be included within the same boundaries, I entitle it the incongruent counterpart of that other. To show its possibility, take a body which is not composed of two halves symmetrically disposed to a single intersecting surface, say a human hand. From all points of its surface draw perpendiculars to a plane set over against it, and produce them just as far behind the plane as these points lie in front of it; the extremities of the lines so produced, if connected, then compose the surface and shape of a physical body which is the incongruent counterpart of the first; i.e., if the given hand is the right, its counterpart is the left. The image of an object in a mirror rests upon the same principle; for it always appears just as far behind the mirror as the object lies in front of its surface, and so the mirrored image of a right hand is always a left. If the object itself consists of two incongruent counterparts, as does the human body when divided by a vertical section from front to back, its image is congruent with it, as can easily be seen by allowing it in thought to make a half turn; for the counterpart of the counterpart of an object is necessarily congruent with the object.

The above considerations may suffice for understanding the possibility of spaces which are completely equal and similar and yet incongruent. We now proceed to the philosophical application of these concepts. From the common example of the two hands, it is already clear that the shape of one body can be completely similar to that of another, and the magnitude of their extension exactly the same, while yet there remains an inner difference, namely that the surface which bounds the one cannot possibly bound the other. Since this surface bounds the physical space of the one but cannot serve as boundary to the other, however one may turn and twist it, this difference must be such as rests upon an inner ground. This inner ground cannot, however, depend on any difference in the mode of connection of the parts of the body relatively to one another; for, as can be seen from the examples adduced, in this respect everything may be completely identical in the two cases. Nevertheless, if we conceive the first created thing to be a human hand, it is necessarily either a right or a left, and to produce the one a different act of the creating cause is required from that whereby its counterpart can come into being.

Should we, then, adopt the conception held by many modern philosophers, especially in Germany, that space consists only in the outer relations of the parts of matter existing alongside one another, in the case before us all actual space would be that which this hand occupies. But since, whether it be right or left, there is no difference in the relations of its parts to one another, the hand would in respect of this characteristic be absolutely indeterminate, i.e., it would fit either side of the human body, which is impossible.

Thus it is evident that instead of the determinations of space following from the positions of the parts of matter relatively to one another, these latter follow from the former. It is also clear that in the constitution of bodies differences are to be found which are real differences, and which are grounded solely in their relation to absolute, primary space. For, only through this relation is the relation of bodily things possible. Since absolute space is not an object of an outer sensation, but a fundamental concept which first makes all such sensations possible, it further follows that whatsoever in the outline of a body exclusively concerns its reference to pure space, can be apprehended only through comparison with other bodies.

A reflective reader will accordingly regard as no mere fiction that concept of space which the geometer has thought out and which clear-thinking philosophers have incorporated into the system of natural philosophy. There is, indeed, no lack of difficulties surrounding this concept, if we attempt to comprehend its reality—a reality which is sufficiently intuitable to inner sense—through ideas of reason. This difficulty always arises when we attempt to philosophise on the first data of our knowledge. But it reaches its maximum when, as in this case, the consequences of an assumed concept [that of spatial relations as subsequent to and dependent on the relations of bodies to one another] contradict the most obvious experience.

Time and Space as Conditions of Knowledge

IMMANUEL KANT, from the *Inaugural Dissertation*

I shall now show that there are two such formal principles of the phenomenal universe which are absolutely primary and universal,[1] and which are, as it were, the schemata and conditions of all human knowledge that is sensitive. I refer to time and space.

14. *On Time*

1. *The idea of time does not originate in the senses, but is presupposed by them.* For the impressions of sense[2] can be represented either as simultaneous or as successive only through the idea of time; succession does not beget the concept of time, but presupposes it. Thus the notion of time (regarded as acquired through experience) is very badly defined in terms of the series of actual things existing *after* one another. For what the word *after* may signify, I know only by means of an antecedently formed concept of time. Things are one after another when they exist at *different times,* just as things are simultaneous when they exist at the *same time.*

2. *The idea of time is singular, not general.* For no time is apprehended except as part of one and the same boundless[3] time. If we think of two years we cannot represent them save by a determinate dating with regard to one another, and if they do not follow one another immediately, save as joined to one another by some intermediate time. But which of different times is earlier, which later, can by no means be defined by any marks

Reprinted by permission of the publisher from *Kant's Inaugural and Early Writings on Space.* Tr. by J. Handyside. The Open Court Publishing Company, La Salle, Illinois, 1929.

conceivable by the intellect, unless we are to incur a vicious circle. The mind does not distinguish earlier and later except by a *singular* intuition. Further, we conceive all actual things as located *in* time, not as contained *under* its general notion as under a common mark.

3. *The idea of time is therefore an intuition.* And since the idea is conceived prior to all sensation, as a condition of relations exhibited in sensibles, the intuition is not sensual but *pure*.

.   .   .   .   .

5. *Time is not something objective and real.* It is neither substance nor accident nor relation, but is a subjective condition, necessary owing to the nature of the human mind, of the co-ordinating of all sensibles according to a fixed law[4]; and it is a *pure intuition.* For we co-ordinate alike substances and accidents, whether according to simultaneity or according to succession, only through the concept of time; and thus the notion of time, as a formal principle, is prior to the concepts of simultaneity and succession. As for relations[5] of whatever sort, in so far as they come within the scope of the senses (namely, as to their simultaneity or succession), they involve nothing further [than] is involved in the co-ordination of substances and [accidents] except this, the determining of positions in time, as either in the same point of it, or in different points.

Those who assert the objective reality of time, conceive it in one or other of two ways. Among English philosophers especially, it is regarded as a continuous real flux, and yet as apart from any existing thing—a most egregious fiction. Leibniz and his School declare it to be a real characteristic, abstracted from the succession of internal states. This latter view at once shows itself erroneous by involving a vicious circle in the definition of time, and also by entirely neglecting simultaneity,[6] a most important consequence of time. It thus upsets the whole use of sane reason, in as much as, instead of requiring the laws of motion to be defined in terms of time, it would have time itself defined in respect of its own nature by reference to the observation of moving things or of some series of internal changes—a procedure by which, clearly, all the certainty of our rules is lost. But as for the fact that we cannot estimate *quantity* of time save in the concrete, namely, either by motion or by the series of [our] thoughts, this is because the concept of time rests only on an internal law of the mind. For since the concept is not a connate intuition, the action of the mind, in the co-ordinating of its sensa, is called forth only by the help of the senses. So far is it from being possible that anyone should ever deduce and explain the concept of time by the help of reason, that the very principle of contradiction presupposes it, involving it as a condition. For A and not-A are not incompatible unless they are judged of the same thing *together* (i.e., in the same time); but when they are judged of a thing successively (i.e., at different times), they may both belong to it. Hence the possibility of changes is thinkable only in time; time is not thinkable through changes, but *vice versa.*

· · · · ·

7. Thus time is an absolutely primary, formal principle of the sensible world. For all things that are in any way sensibles can be apprehended only as at the same time or in successive times, and so as included and definitely related to each other within the course of the one single time. Thus through this concept, primary in the domain of sense, there necessarily arises a formal whole which is not a part of any other, i.e., the phenomenal world.

· · · · ·

15. On Space

A. *The concept of space is not abstracted from outer sensations.* For I cannot conceive anything as located outside me unless I represent it as in a space different from the space in which I myself am; nor can I conceive things as outside one another unless I arrange them in different parts of space. Therefore the possibility of outer perceptions, as such, presupposes, and does not create, the concept of space. Inasmuch as, moreover, things which are in space [and which presuppose space] affect the senses, space itself [on which the apprehension of them depends] cannot be derived from the senses.

B. *The concept of space is a singular representation,* including all spaces *in* itself, not an abstract common notion containing them *under* itself. For what we call "many spaces" are only parts of the same boundless space, with a certain position relatively to one another; nor can we conceive a cubic foot except as coterminous on all sides with surrounding space.

C. *The concept of space is thus a pure intuition,* since it is a singular concept. It is not put together from sensations, but is the fundamental form of all outer sensation. This pure intuition can be readily observed in the axioms of geometry, and in every mental construction of postulates or of problems. For that space has not more than three dimensions, that there is but one straight line between two points, that from a given point in a plane surface with a given straight line as radius a circle can be described, etc., are not inferred from any universal notion of space, but can only be discerned[7] in space in the concrete. We cannot by any sharpness of intellect describe discursively, that is, by intellectual marks, the distinction in a given space between things which lie towards one quarter, and things which are turned towards the opposite quarter. Thus if we take solids completely equal and similar but incongruent, such as the right and left hands (so far as they are conceived only according to extension), or spherical triangles from two opposite hemispheres, although in every respect which admits of being stated in terms intelligible to the mind through a verbal description they can be substituted for one another, there is yet a diversity which makes it impossible for the boundaries of extension to coincide. It is therefore clear that in these cases the diversity, that is, the incongruence, cannot be apprehended except by pure intuition. Hence geometry employs principles which not only are unquestioned and discursive, but which are such as fall under the mind's direct observation. *Evidence* in demonstrations (meaning thereby the *clearness* of assured knowledge, so far as this clearness can be likened to that of sense) is found in geometry in not merely in the highest degree, but is found there alone of all the pure sciences. Geometrical evidence is thus the model for, and the means of attaining, all evidence in the other sciences. For since geometry contemplates the relations of space, the concept of which contains in itself the very form of all sensual intuition, there can be nothing clear and perspicuous in things perceived by outer sense except through the mediation of the intuition which that science is occupied in contemplating. Further, geometry does not demonstrate its universal propositions by apprehending the object through a universal concept, as is done in matters of reason, but by submitting it to the eyes in a singular intuition, as is done in matters of sense.[8]

D. *Space is not something objective and real,* neither substance, nor accident, nor relation, *but subjective and ideal;* and, as it were, a schema, issuing by a constant law from the nature of the mind, for the co-ordinating of all outer sensa whatsoever. Those who defend the reality of space, either conceive it as an absolute and boundless receptacle of possible things (the view commends itself to most geometers, following the English), or hold that it is itself a relation of existent things, vanishing therefore if things be annihilated, and not thinkable except in actual things (as, following Leibniz, most of our countrymen maintain). The former empty figment of reason, since it imagines an infinity of real relations without any things which are so related, pertains to the world of fable. But those who adopt the second opinion fall into a much more serious error. They dash down geometry from the supreme height of certainty, reducing it to the rank of those sciences whose principles are empirical. For if all properties of space are borrowed only from external relations through experience, geometrical axioms do not possess universality, but only that comparative universality which is acquired through induction and holds only so widely as it is observed; nor do they possess necessity, except such as depends on fixed laws of nature; nor have they any precision save such as is matter of arbitrary convention; and we might hope, as in empirical matters, some day to discover a space endowed with other primary affections, and perhaps even a rectilinear figure enclosed by two straight lines.

For things cannot appear to the senses in any manner[9] except by the mediating power of the mind, co-ordinating all sensations according to a constant law inborn in it. Nothing whatsoever, then, can be given to the senses save in conformity with the primary axioms of space and the other consequences of its nature, as expounded by geometry. Though the ground or principle of these axioms be only subjective, it will necessarily be in harmony with them, because only so does it harmonise with itself. The laws of sensibility will be laws of nature, in so far as nature is able to affect the senses. Thus as regards all properties of space which are demonstrated from a hypothesis not invented, but intuitively given as being a subjective condition of all phenomena, nature is meticulously conformed to the rules of geometry, and only in accordance with them can nature be revealed to the senses.

For only if infinite space and time be given, can any definite space or time be marked out by limitation of it; neither a point nor a moment can be thought by itself; they are conceived only as limits in an already given space or time. Thus all primary properties of these concepts are beyond the jurisdiction of reason, and so cannot in any way be intellectually explained. But none the less they are the pre-suppositions upon which the intellect rests when, with the greatest possible certainty, and in accordance with logical laws, it draws consequences from the primary data of intuition.

Finally, the question naturally arises, whether these concepts [space and time] are connate or acquired. The latter alternative, it is true, seems already refuted by our demonstrations, but the former is not to be rashly admitted, since, in appealing to a first cause, it opens a path for that lazy philosophy which declares all further research to be vain. Both concepts are without doubt acquired, as abstracted, not indeed from the sensing of objects[10] (for sensation gives the matter, not the form, of human apprehension), but from the action of the mind in co-ordinating its sensa according to unchanging[11] laws—each being, as it were, an immutable type, and therefore to be known intuitively. For, though sensations excite this act of the mind, they do not determine the intuition. Nothing is here connate save the law of the mind, according to which it combines in a fixed manner the sensa produced in it by the presence of the object.

On the Law of Inertia

ERNST MACH, from *The Science of Mechanics*

According to Einstein, the work of the Austrian physicist and philosopher Ernst Mach (1838–1916) had a strong influence upon the direction of his investigations. Mach believed that the tradition of Newton and Euler in mechanics contained much that was physically true, but it also contained much that was philosophically untenable. This belief made it impossible for him to adopt their views literally, and he felt that the truths of that tradition had to be reformulated in a cogent manner.

For example, Newton believed that all bodies have mass. But this implied (for Newton) that they have a certain quantity of matter. According to Mach, this matter is something which can never be perceived. Matter is that something which has properties. Thus it can be red, but the only things to be seen are the red things, not the matter itself. Matter underlies what we see, but it is something which is never seen. For this reason Mach asserted that matter as well as Absolute Space and Absolute Time were scientifically worthless, physically meaningless, illegitimate concepts of science.

Newton seems to have been just as aware as Mach that Absolute Space could not be perceived (Newton, p. 64), and he offered what he believed to be evidence for its existence—one part of which was his bucket experiment. This experiment, it will be recalled, consists of a bucket of water suspended by a twisted string from some fixed hook. The bucket is released and begins to spin. Initially (stage 1) the level of the water in the bucket is flat. As the bucket spins more rapidly (stage 2) some of the motion is communicated to the water, and the level of water changes shape, having now a dip in it which reaches up the sides of the bucket. As the bucket spins (stage 3), the bucket and the water in it spin at the same rate, and the water still has the dip in it. What caused the level of the water to dip? According to

Reprinted by permission of the publisher from *The Science of Mechanics, A Critical and Historical Account of its Development.* Tr. by T. J. McCormack with new introduction by K. Menger. The Open Court Publishing Company, La Salle, Illinois, 1960. Original German edition, 1883.

Newton, this change cannot be explained as due to the rotation of the water with respect to any body; therefore he offered the experiment as evidence for rotation with respect to Absolute Space.

Mach pointed out in his equally famous antibucket argument that Newton had committed a *non sequitur*. The experiment shows that it is not rotation with respect to the bucket which causes dips. For the dip occurs even in stage 3, when there is no relative rotation between the water and the bucket. But that simply rules out the bucket. The dipping effect might still be due to rotation of the water with respect to some other body. In particular, Mach wanted to argue that it was rotation with respect to the fixed stars which caused the depression to occur. Mach's counterargument won the day and neutralized the relevance of the bucket experiment. This was a pity because, in our opinion, Mach misunderstood Newton's argument.

Newton was trying to refute a theory of mechanics held by Descartes and his followers. According to the Cartesians, if there was a change in a body, then it was due to the actions of other bodies in its immediate neighborhood. In Newton's experiment, the bucket is the only body immediately near the depressed body of water. Therefore it was exactly to the point to show, as Newton did, that motion with respect to the bucket had no causal efficacy. Newton was thus offering evidence to show that his own theory was preferable to Descartes'.

Mach felt, for these various reasons, that he had to reformulate the Law of Inertia as well as the rest of mechanics so that it no longer employed illegitimate concepts nor rested upon apparently spurious evidence. As part of the program of reform, Mach proposed that we consider the ratio of masses of bodies as a function of the accelerations which two bodies induce in each other. He then argued that there were certain truths or theorems of experience, as he called them, on the basis of which he could show that the notion of mass he had introduced fulfilled all the functions which Newton required of the concept. Moreover, we would now possess, if Mach was correct, a concept of mass which did not rest upon the physically meaningless concept of matter. It was a brilliant analysis which was but one part of a widely conceived program and which is widely employed today.

Another kind of consideration moved Mach to modify the laws of mechanics. It is no less conceptual than the other considerations we have mentioned. But in tone it is very like the motivation of Einstein (see Einstein, p. 155), and it leads to a problem which has become known as Mach's Problem, or Mach's Thesis.

According to Mach, whether we describe the Earth as being at rest and the fixed stars rotating, or the stars at rest and the Earth rotating, is geometrically exactly the same case, though one description is astronomically more simple. But the two descriptions are not interchangeable with respect to explanations. For the Law of Inertia can explain why the Earth is flattened at the poles if it is the Earth which rotates. But the Law cannot explain the flattening if it is the fixed stars which rotate and the Earth is at rest. There are, according to Mach, two solutions to this problem. In the first, Absolute Space is assumed to exist, in which case the two descriptions are not interchangeable—one of them is false. According to the second solution, it is admitted that the Law of Inertia has been wrongly stated, and it is reformulated so that it has the same explanatory power in either of the presumably equivalent cases.

Mach thought that in order for the modified Law of Inertia to have wide application, the inertial motion of a body should be described with respect not to just a few key bodies, but with respect to all the masses of the universe. Paradoxically, the frame of reference which Mach found most practical is impossible to identify in practice, since it refers to all the masses of the universe. We know that a body moving inertially will move with a constant speed in a certain direction. This leads us to Mach's related question: What influence do all the bodies of the universe have in determining the direction and velocity referred to in the Law of Inertia? This is still an open question today (see Sciama, p. 185).

"The natural days, which, commonly, for the purpose of the measurement of time, are held as equal, are in reality unequal. Astronomers correct this inequality, in order that they may measure by a truer time the celestial motions. It may be that there is no equable motion, by which time can accurately be measured. All motions can be accelerated and retarded. But the flow of *absolute* time cannot be changed. Duration, or the persistent existence of things, is always the same, whether motions be swift or slow or null."*

2. It would appear as though Newton in the remarks here cited still stood under the influence of the mediæval philosophy, as

---

* [For the expanded version of this extrapolation from Newton's "Scholium," Art. IV, see pp. 74–75—Ed.]

though he had grown unfaithful to his resolves to investigate only actual facts. When we say a thing *A* changes with the time, we mean simply that the conditions that determine a thing *A* depend on the conditions that determine another thing *B*. The vibrations of a pendulum take place *in time* when its excursion *depends* on the position of the earth. Since, however, in the observation of the pendulum, we are not under the necessity of taking into account its dependence on the position of the earth, but may compare it with any other thing (the conditions of which of course also depend on the position of the earth), the illusory notion easily arises that *all* the things with which we compare it are unessential. Nay, we may, in attending to the motion of a pendulum, neglect entirely other external things, and find that for every position of it our thoughts and sensations are different. Time, accordingly, appears to be some particular and independent thing, on the progress of which the position of the pendulum depends, while the things that we resort to for comparison and choose at random appear to play a wholly collateral part. But we must not forget that all things in the world are connected with one another and depend on one another, and that we ourselves and all our thoughts are also a part of nature. It is utterly beyond our power to *measure* the changes of things by *time*. Quite the contrary, time is an abstraction, at which we arrive by means of the changes of things; made because we are not restricted to any one *definite* measure, all being interconnected. A motion is termed uniform in which equal increments of space described correspond to equal increments of space described by some motion with which we form a comparison, as the rotation of the earth. A motion may, with respect to another motion, be uniform. But the question whether a motion is *in itself* uniform, is senseless. With just as little justice, also, may we speak of an "absolute time"—*of a time independent of* change. This absolute time can be measured by comparison with no motion; it has therefore neither a practical nor a scientific value; and no one is justified in saying that he knows aught about it. It is an idle metaphysical conception.

.　　.　　.　　.　　.

We arrive at the idea of time—to express it briefly and popularly—by the connection of that which is contained in the province of our memory with that which is contained in the province of our sense-perception. When we say that time flows on in a definite direction or sense, we mean that physical events generally (and therefore also physiological events) take place only in a definite sense.[1] Differences of temperature, electrical differences, differences of level generally, if left to themselves, all grow less and not greater. If we contemplate two bodies of different temperatures, put in contact and left wholly to themselves, we shall find that it is possible only for greater differences of temperature in the field of memory to exist with lesser ones in the field of sense-perception, and not the reverse. In all this there is simply expressed a peculiar and profound connection of things. To demand at the present time a full elucidation of this matter, is to anticipate, in the manner of speculative philosophy, the results of all future special investigation, that is, a perfected physical science.

. . . *All* masses and *all* velocities, and consequently *all* forces, are relative. There is no decision about relative and absolute which we can possibly meet, to which we are forced, or from which we can obtain any intellectual or other advantage. When quite modern authors let themselves be led astray by the Newtonian arguments which are derived from the bucket of water, to distinguish between relative and absolute motion, they do not reflect that the system of the world is only given *once* to us, and the Ptolemaic or Copernican view is *our* interpretation, but both are equally actual. Try to fix Newton's bucket and rotate the heaven of fixed stars and then prove the absence of centrifugal forces.

4. It is scarcely necessary to remark that in the reflections here presented Newton has again acted contrary to his expressed intention only to investigate *actual facts*. No one is competent to predicate things about absolute space and absolute motion; they are pure things of thought, pure mental constructs, that cannot be produced in experience. All our principles of mechanics are, as we have shown in detail, experimental knowledge concerning the relative positions and motions of bodies. Even in the provinces in which they are now recognized as valid, they could not, and were not, admitted without previously being subjected to experimental tests. No one is warranted in extending these principles beyond the boundaries of experience. In fact, such an extension is meaningless, as no one possesses the requisite knowledge to make use of it.

We must suppose that the change in the point of view from which the system of the world is regarded which was initiated by Copernicus, left deep traces in the thought of Galileo and Newton. But while Galileo, in his theory of the tides, quite naïvely chose the sphere of the fixed stars as the basis of a new system of coördinates, we see doubts expressed by Newton as to whether a given fixed star is at rest only apparently or really (*Principia,* 1687, p. 11). This appeared to him to cause the difficulty of distinguishing between true (absolute) and apparent (relative) motion. By this he was also impelled to set up the conception of *absolute space.* By further investigations in this direction—the discussion of the experiment of the rotating spheres which are connected together by a cord and that of the rotating water-bucket (pp. 9, 11)—he believed that he could prove an absolute rotation, though he could not prove any absolute translation. By absolute rotation he understood a rotation relative to the fixed stars, and here centrifugal forces can always be found. "But how we are to collect," says Newton in the Scholium at the end of the Definitions, "the true motions from their causes, effects, and apparent differences, and *vice versa;* how from the motions, either true or apparent, we may come to the knowledge of their causes and effects, shall be explained more at large in the following Tract." The resting sphere of fixed stars seems to have made a certain impression on Newton as well. The natural system of reference is for him that which has any uniform motion or translation without rotation (relatively to the sphere of fixed stars).[2] But do not the words quoted in inverted commas give the impression that Newton was glad to be able now to pass over to less precarious questions that could be tested by experience?

Let us look at the matter in detail. When we say that a body K alters its direction and velocity solely through the influence of another body K', we have asserted a conception that it is impossible to come at unless other bodies $A, B, C \ldots$ are present with reference to which the motion of the body K has been estimated. In reality, therefore, we are simply cognizant of a relation of the body K to $A, B, C. \ldots$ If now we suddenly neglect $A, B, C \ldots$ and attempt to speak of the deportment of the body K in absolute space, we implicate ourselves in a twofold error. In the first place, we cannot know how K would act in the absence of $A, B, C \ldots$; and in the second place, every means would be wanting of forming a judgment of the behavior of K and of putting to the test what

we had predicated—which latter therefore would be bereft of all
scientific significance.

. . . . .

5. Let us now examine the point on which Newton, apparently
with sound reasons, rests his distinction of absolute and relative
motion. If the earth is affected with an *absolute* rotation about
its axis, centrifugal forces are set up in the earth: it assumes an
oblate form, the acceleration of gravity is diminished at the
equator, the plane of Foucault's pendulum rotates, and so on. All
these phenomena disappear if the earth is at rest and the other
heavenly bodies are affected with absolute motion round it, such
that the same *relative* rotation is produced. This is, indeed, the
case, if we start *ab initio* from the idea of absolute space. But if
we take our stand on the basis of facts, we shall find we have
knowledge only of *relative* spaces and motions. *Relatively*, not
considering the unknown and neglected medium of space, the
motions of the universe are the same whether we adopt the
Ptolemaic or the Copernican mode of view. Both views are, indeed, equally *correct;* only the latter is more simple and more
practical. The universe is not *twice* given, with an earth at rest
and an earth in motion; but only *once,* with its *relative* motions,
alone determinable. It is, accordingly, not permitted us to say
how things would be if the earth did not rotate. We may interpret
the one case that is given us, in different ways. If, however, we
so interpret it that we come into conflict with experience, our
interpretation is simply wrong. The principles of mechanics
can, indeed, be so conceived, that even for relative rotations
centrifugal forces arise.

Newton's experiment with the rotating vessel of water simply
informs us that the relative rotation of the water with respect
to the sides of the vessel produces *no* noticeable centrifugal forces,
but that such forces *are* produced by its relative rotation with
respect to the mass of the earth and the other celestial bodies.
No one is competent to say how the experiment would turn out if
the sides of the vessel increased in thickness and mass till they
were ultimately several leagues thick. The one experiment only
lies before us, and our business is to bring it into accord with the
other facts known to us, and not with the arbitrary fictions of our
imagination.

6. When Newton examined the principles of mechanics dis-

covered by Galileo, the great value of the simple and precise law of inertia for deductive derivations could not possibly escape him. He could not think of renouncing its help. But the law of inertia, referred in such a naïve way to the earth supposed to be at rest, could not be accepted by him. For, in Newton's case, the rotation of the earth was not a debatable point; it rotated without the least doubt. Galileo's happy discovery could only hold approximately for small times and spaces, during which the rotation did not come into question. Instead of that, Newton's conclusions about planetary motion, referred as they were to the fixed stars, appeared to conform to the law of inertia. Now, in order to have a generally valid system of reference, Newton ventured the fifth corollary of the *Principia* (p. 19 of the first edition).[3] He imagined a momentary terrestrial system of coördinates, for which the law of inertia is valid, held fast in space without any rotation relatively to the fixed stars. Indeed he could, without interfering with its usability, impart to this system any initial position and any uniform translation relatively to the above momentary terrestrial system. The Newtonian laws of force are not altered thereby; only the initial positions and initial velocities—the constants of integration—may alter. By this view Newton gave the *exact* meaning of his hypothetical extension of Galileo's law of inertia. We see that the reduction to absolute space was by no means necessary, for the system of reference is just as relatively determined as in every other case. In spite of his metaphysical liking for the absolute, Newton was correctly led by the *tact of the natural investigator*. This is particularly to be noticed, since, in former editions of this book, it was not sufficiently emphasized. How far and how accurately the conjecture will hold good in future is of course undecided.

The comportment of terrestrial bodies with respect to the earth is reducible to the comportment of the earth with respect to the remote heavenly bodies. If we were to assert that we knew more of moving objects than this their last-mentioned, experimentally-given comportment with respect to the celestial bodies, we should render ourselves culpable of a falsity. When, accordingly, we say that a body preserves unchanged its direction and velocity *in space,* our assertion is nothing more or less than an abbreviated reference to *the entire universe.* The use of such an abbreviated expression is permitted the original author of the principle, because he knows that as things are, no difficulties stand in the way of carrying out its implied directions. But no remedy lies in his power, if difficulties of the kind mentioned present themselves; if, for example, the requisite, relatively fixed bodies are wanting.

7. Instead, now, of referring a moving body K to space, that is to say to a system of coördinates, let us view directly its relation to the bodies of the universe, by which alone such a system of coördinates can be determined.

.   .   .   .   .

9. We have attempted in the foregoing to give the law of inertia a different expression from that in ordinary use. This expression will, so long as a sufficient number of bodies are apparently fixed in space, accomplish the same as the ordinary one. It is as easily applied, and it encounters the same difficulties. In the one case we are unable to come at an absolute space, in the other a limited number of masses only is within the reach of our knowledge, and the summation indicated can consequently not be fully carried out. It is impossible to say whether the new expression would still represent the true conditions of things if the stars were to perform rapid movements among one another. The general experience cannot be constructed from the particular case given us. We must, on the contrary, *wait* until such an experience presents itself. Perhaps when our physico-astronomical knowledge has been extended, it will be offered somewhere in celestial space, where more violent and complicated motions take place than in our environment. The most important result of our reflections is, however, *that precisely the apparently simplest mechanical principles are of a very complicated character, that these principles are founded on uncompleted experiences, nay on experiences that never can be fully completed, that practically, indeed, they are sufficiently secured, in view of the tolerable stability of our environment, to serve as the foundation of mathematical deduction, but that they can by no means themselves be regarded as mathematically established truths but only as principles that not only admit of constant control by experience but actually require it.*

The Definition of Mass

ERNST MACH, from *History and Root of the Principle of the Conservation of Energy*

Obviously it does not matter whether we think of the earth as turning round on its axis, or at rest while the celestial bodies revolve round it. Geometrically these are exactly the same case of a relative rotation of the earth and of the celestial bodies with respect to one another. Only, the first representation is astronomically more convenient and simpler.

But if we think of the earth at rest and the other celestial bodies revolving round it, there is no flattening of the earth, no Foucault's experiment,[1] and so on—at least according to our usual conception of the law of inertia. Now, one can solve the difficulty in two ways: Either all motion is absolute, or our law of inertia is wrongly expressed. Neumann preferred the first supposition; I, the second. The law of inertia must be so conceived that exactly the same thing results from the second supposition as from the first. By this it will be evident that, in its expression, regard must be paid to the masses of the universe.

In ordinary terrestrial cases, it will answer our purposes quite well to reckon the direction and velocity with respect to the top of a tower or a corner of a room; in ordinary astronomical cases, one or other of the stars will suffice. But because we can also choose other corners of rooms, another pinnacle, or other stars, the view may easily arise that we do not need such a point at all from which to reckon. But this is a mistake; such a system of co-ordinates has a value only if it can be determined by means of bodies. We here fall into the same error as we did with the representation of time. Because a piece of paper money need not neces-

Reprinted, by permission of the publisher, from *History and Root of the Principle of the Conservation of Energy*. Tr. and annotated by P. E. B. Jourdain. The Open Court Publishing Company, La Salle, Illinois, 1911.

sarily be funded by a definite piece of money, we must not think that it need not be funded at all.

In fact, any one of the above points of origin of co-ordinates answers our purposes as long as a sufficient number of bodies keep fixed positions with respect to one another. But if we wish to apply the law of inertia in an earthquake, the terrestrial points of reference would leave us in the lurch, and, convinced of their uselessness, we would grope after celestial ones. But, with these better ones, the same thing would happen as soon as the stars showed movements which were very noticeable. When the variations of the positions of the fixed stars with respect to one another cannot be disregarded, the laying down of a system of co-ordinates has reached an end. It ceases to be immaterial whether we take this or that star as point of reference; and we can no longer reduce these systems to one another. We ask for the first time which star we are to choose, and in this case easily see that the stars cannot be treated indifferently, but that because we can give preference to none, the influence of all must be taken into consideration.

We can, in the application of the law of inertia, disregard any particular body, provided that we have enough other bodies which are fixed with respect to one another. If a tower falls, this does not matter to us; we have others. If Sirius alone, like a shooting-star, shot through the heavens, it would not disturb us very much; other stars would be there. But what would become of the law of inertia if the whole of the heavens began to move and the stars swarmed in confusion? How would we apply it then? How would it have to be expressed then? We do not inquire after one body as long as we have others enough; nor after one piece of money as long as we have others enough. Only in the case of a shattering of the universe, or a bankruptcy, as the case may be, we learn that *all* bodies, each with its share, are of importance in the law of inertia, and all money, when paper money is funded, is of importance, each piece having its share.

.　.　.　.　.

Now, what share has every mass in the determination of direction and velocity in the law of inertia? No definite answer can be given to this by our experiences. We only know that the share of the nearest masses vanishes in comparison with that of the farthest.

We would, then, be able completely to make out the facts known to us if, for example, we were to make the simple supposition that all bodies act in the way of determination proportionately to their masses and independently of the distance, or proportionately to the distance, and so on. Another expression would be: In so far as bodies are so distant from one another that they contribute no noticeable acceleration to one another, all distances vary proportionately to one another.

.   .   .   .   .

On the Definition of Mass[2]

The circumstance that the fundamental propositions of mechanics are neither wholly *a priori* nor can wholly be discovered by means of experience—for sufficiently numerous and accurate experiments cannot be made—results in a peculiarly inaccurate and unscientific treatment of these fundamental propositions and conceptions. Rarely is distinguished and stated clearly enough what is *a priori,* what empirical, and what is hypothesis.

Now, I can only imagine a scientific exposition of the fundamental propositions of mechanics to be such that one regards these theorems as hypotheses to which experience forces us, and that one afterwards shows how the denial of these hypotheses would lead to contradictions with the best-established facts.

As evident *a priori* we can only, in scientific investigations, consider the law of causality or the law of sufficient reason, which is only another form of the law of causality. No investigator of nature doubts that under the same circumstances the same always results, or that the effect is completely determined by the cause. It may remain undecided whether the law of causality rests on a powerful induction or has its foundation in the psychical organization (because in the psychic life, too, equal circumstances have equal consequences).

The importance of the law of sufficient reason in the hands of an investigator was proved by Clausius's works on thermodynamics and Kirchhoff's researches on the connexion of absorption and emission. The well-trained investigator accustoms himself in his thought, by the aid of this theorem, to the same definiteness as nature has in its actions, and then experiences which are not in themselves very apparent suffice, by exclusion of all that is contradictory, to discover very important laws connected with the said experiences.

Usually, now, people are not very chary of asserting that a proposition is immediately evident. For example, the law of inertia is often stated to be such a proposition, as if it did not need the proof of experience. The fact is that it can only have grown out of experience. If masses imparted to one another, not acceleration, but, say, velocities which depended on the distance, there would be no law of inertia; but whether we have the one state of things or the other, only experience teaches. If we had merely sensations of heat, there would be merely equalizing velocities *(Ausgleichungsgeschwindigkeiten)*, which vanish with the differences of temperature.

One can say of the motion of masses: "The effect of every cause persists," just as correctly as the opposite: "Cessante causa cessat effectus"; it is merely a matter of words. If we call the resulting velocity the "effect," the first proposition is true, if we call the acceleration the "effect," the second is true.

Also people try to deduce *a priori* the theorem of the parallelogram of forces; but they must always bring in tacitly the supposition that the forces are independent of one another. But by this the whole derivation becomes superfluous.

I will now illustrate what I have said by *one* example, and show how I think the conception of mass can be quite scientifically developed. The difficulty of this conception, which is pretty generally felt, lies, it seems to me, in two circumstances: (1) in the unsuitable arrangement of the first conceptions and theorems of mechanics; (2) in the silent passing over important presuppositions lying at the basis of the deduction.

Usually people define $m = p/g$ and again $p = mg$. This is either a very repugnant circle, or it is necessary for one to conceive force as "pressure." The latter cannot be avoided if, as is customary, statics precedes dynamics. The difficulty, in this case, of defining magnitude and direction of a force is well known.

In that principle of Newton, which is usually placed at the head of mechanics, and which runs: "Actioni contrariam semper et aequalem esse reactionem: sive corporum duorum actiones in se mutuo semper esse aequales et in partes contrarias dirigi," the "actio" is again a pressure, or the principle is quite unintelligible unless we possess already the conception of force and mass. But pressure looks very strange at the head of the quite phoronomical mechanics of today. However, this can be avoided.

If there were only one kind of matter, the law of sufficient reason would be sufficient to enable us to perceive that two completely similar bodies can impart to each other only *equal* and *opposite* accelerations. This is the one and only effect which is completely determined by the cause.

Now, if we suppose the mutual independence of forces, the following easily results. A body A, consisting of m bodies a, is the presence of another body B, consisting of m' bodies a. Let the acceleration of A be ϕ and that of B be ϕ'. Then we have $\phi:\phi' = m':m$.

If we say that a body A has the mass m if it contains the body a m times, this means that the accelerations vary as the masses.

To find by experiment the mass-ratio of two bodies, let us allow them to act on one another, and we get, when we pay attention to the sign of the acceleration, $m/m' = -(\phi'/\phi)$.

If the one body is taken as a unit of mass, the calculation gives the mass of the other body. Now, nothing prevents us from applying this definition in cases in which two bodies of different matter act on one another. Only, we cannot know *a priori* whether we do not obtain other values for a mass when we consult other bodies used for purposes of comparison and other forces. When it was found that A and B combine chemically in the ratio $a:b$ of their weights and that A and C do so in the ratio $a:c$ of their weights, it could not be known beforehand that B and C combine in the ratio $b:c$. Only experience can teach us that two bodies which behave to a third as equal masses will also behave to one another as equal masses.

If a piece of gold is opposed to a piece of lead, the law of sufficient reason leaves us completely. We are not even justified in expecting contrary motions: both bodies might accelerate in the same direction. The calculation would then lead to negative masses.

But that two bodies which behave as equal masses to a third behave as such to one another, with respect to any forces, is very likely, because the contrary would not be reconcilable with the law of the conservation of work (*Kraft*), which has hitherto been found to be valid.

Imagine three bodies A, B, and C movable on an absolutely smooth and absolutely fixed ring. The bodies are to act on one another with any forces. Further, both *A* and *B* [Fig. 1], on the one hand, and *A* and *C*, on the other, are to behave to one another as equal masses. Then the same must hold between *B* and *C*.

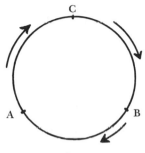

Fig. 1

If, for example, *C* behaved to *B* as a greater mass to a lesser one, and we gave *B* a velocity in the direction of the arrow, it would give this velocity wholly to *A* by impact, and *A* would give it wholly to *C*. Then *C* would communicate to *B* a greater velocity and yet keep some itself. With every revolution in the direction of the arrow, then, the *vis viva* in the ring would increase; and the contrary would take place if the original motion were in a direction opposite to that of the arrow. But this would be in glaring contradiction with the facts hitherto known.

If we have thus defined mass, nothing prevents us from keeping the old definition of force as product of mass and acceleration. The law of Newton[3] mentioned above then becomes a mere identity.

Since all bodies receive from the earth an equal acceleration, we have in this force (their weight) a convenient measure of their masses; again, however, only under the two suppositions that bodies which behave as equal masses to the earth do so to one another, and with respect to every force. Consequently, the following arrangement of the theorems of mechanics would appear to me to be the most scientific.

THEOREM OF EXPERIENCE

Bodies placed opposite to one another communicate to each other accelerations in opposite senses in the direction of their line of junction. The law of inertia is included in this.

DEFINITION

Bodies which communicate to each other equal and opposite accelerations are said to be of equal mass. We get the mass-value of a body by dividing the acceleration which it gives the body with which we compare others, and choose as the unit, by the acceleration which it gets itself.

THEOREM OF EXPERIENCE

The mass-values remain unaltered when they are determined with reference to other forces and to another body of comparison which behaves to the first one as an equal mass.

THEOREM OF EXPERIENCE

The accelerations which many masses communicate to one another are mutually independent. The theorem of the parallelogram of forces is included in this.

DEFINITION

Force is the product of the mass-value of a body into the acceleration communicated to that body.

The Simultaneity of Events

ALBERT EINSTEIN, from *On the Electrodynamics of Moving Bodies*

The first paper of Albert Einstein (1879–1955) on the Special
Theory of Relativity is rightly acknowledged as a turning point
in man's thinking about the physical world. His paper began in a
very deceptive way by calling attention to certain asymmetries between electromagnetic theory and phenomena. Einstein noted that
if there were a sense to Absolute Rest, then there would be a point
in distinguishing between one body being at rest and a second
moving uniformly with respect to it, or the other way around. But
attempts to detect whether bodies are at rest or in uniform motion
yielded no positive answer. If the difference between two states of
motion cannot be detected, then something is wrong with a theory
which tries to explain effects by referring to such undetectable
differences. Einstein suggested that Maxwell's theory should be
modified so that it does not matter which description of motion is
used: the predictions and explanations of the modified theory
should be the same upon either description.

Thus, it looked as if Einstein set out to repair a conceptual difficulty in Maxwell's theory. He sought an electrodynamic theory
whose explanations did not rely on the fact that a body was at
rest or in uniform motion, but referred instead to the relative
motion of the two bodies. Einstein demanded this not only of
any satisfactory theory of electrodynamics—he also made it a requirement of any theory whatever. Let us suppose that a law of
nature describes the behavior of physical systems and how
such systems develop in time. We shall suppose further that this
development or change can be specified by something called "the
state of the system" so that a law will tell us about a temporal
sequence of states. Laws may tell us more than this, but they will
tell us at least this. Einstein's Principle of Relativity states that

Reprinted by permission of the publisher from *The Principle of Relativity*.
Tr. by W. Perrett and G. B. Berry,[1] with notes by A. Sommerfeld. Dover Pub-
lications, New York, 1923.

"the laws by which the states of physical systems undergo change are not affected, whether those changes of state be referred to the one or the other of two systems of coordinates in uniform translatory motion." Einstein required that all sciences contain only those laws which conform to the Principle of Relativity. If a science like mechanics employed a statement which did not conform, then the statement had to be replaced by another which met the relativistic conditions. The goal is a relativistic mechanics whose laws have exactly the same form for all observers in uniform motion with respect to each other. Einstein emphasized the importance of those structures which are the same for all observers, so that the term "relativity" is a little misleading in the present context.

The methodological principle embodied in his Principle is but one of the major insights closely associated with Einstein. Another is his analysis of the concept of simultaneity, which led him to discover new non-Newtonian concepts of space and time, and mass, which now bear his name and which play such prominent roles in modern theories of matter. Einstein believed that our judgments about the temporal order of events are based upon judgments of simultaneity; that is, the occurrence of two events at the same time. His contention was that we make such judgments about events which occur near each other. But to say that two events took place at the same time, when they did not take place near one another, is a very different kind of judgment. Einstein argued that in order to compare the times of two distant events, a *definition* of "simultaneity," or "synchronous," is needed. The one he suggested seems a natural one. A light ray is sent from one place to another and is immediately reflected back. If the transit time (measured by the difference between the clocks at each place) is the same for the going and the return of the ray, then the two clocks are called synchronous.

This analysis of the simultaneity of two events was shown to lead to a rather startling consequence when conjoined with the second postulate of his Theory of Special Relativity: the velocity of a ray of light (*in vacuo*) is independent of whether the ray is emitted by a stationary or moving body. Although two events might be simultaneous when referred to one coordinate system, they would in general fail to be simultaneous when referred to a system moving uniformly with respect to the first. In other words, we cannot speak simply of two events being simultaneous; we must take note of the observer's state of motion. Nor is that the end of the matter. The most basic quantities of physics were soon shown also to depend upon the velocity of the observer or the system to which the physical quantities were referred. Thus, a fundamental concept such as the length of a body became the subject of spirited debates. If one measured the length of a rod which was at rest, and then measured it with the aid of a photometric device as it moved by, the results differed. Scientists began to talk of the true or real versus the apparent length of the rod, and they wondered which length was the real one: the one measured when the rod was at rest or when in motion? Conceptual changes were also introduced in the concepts of time and mass.

In the second selection, given here, Einstein described the empirical and conceptual ferment which he helped to introduce. In it he reflects upon those exciting changes in our concept of space and time required by his second relativity theory, the General Theory of Relativity, and he sets in relief those key features of space and time which have remained the same despite all the change.

It is known that Maxwell's electrodynamics—as usually understood at the present time—when applied to moving bodies, leads to asymmetries which do not appear to be inherent in the phenomena. Take, for example, the reciprocal electrodynamic action of a magnet and a conductor. The observable pheonmenon here depends only on the relative motion of the conductor and the magnet, whereas the customary view draws a sharp distinction between the two cases in which either the one or the other of these bodies is in motion. For if the magnet is in motion and the conductor at rest, there arises in the neighbourhood of the magnet an electric field with a certain definite energy, producing a current at the places where parts of the conductor are situated. But if the magnet is stationary and the conductor in motion, no electric field arises in the neighbourhood of the magnet. In the conductor, however, we find an electromotive force, to which in itself there is no corresponding energy, but which gives rise—assuming equality of relative motion in the two cases discussed—to electric currents of the same path and intensity as those produced by the electric forces in the former case.

Examples of this sort, together with the unsuccessful attempts to discover any motion of the earth relatively to the "light medium," suggest that the phenomena of electrodynamics as well as of mechanics possess no properties corresponding to the idea of absolute rest. They suggest rather that, as has already been shown to the first order of small quantities, the same laws of electrodynamics and optics will be valid for all frames of reference for which the equations of mechanics hold good.[2] We will raise this conjecture (the purport of which will hereafter be called the "Principle of Relativity") to the status of a postulate, and also introduce another postulate, which is only apparently irreconcilable with the former, namely, that light is always propagated in empty space with a definite velocity c which is independent of the state of motion of the emitting body. These two postulates suffice for the attainment of a simple and consistent theory of the electrodynamics of moving bodies based on Maxwell's theory for stationary bodies. The introduction of a "luminiferous ether" will prove to be superfluous inasmuch as the view here to be developed will not require an "absolutely stationary space" provided with special properties, nor assign a velocity-vector to a point of the empty space in which electromagnetic processes take place.

The theory to be developed is based—like all electrodynamics —on the kinematics of the rigid body, since the assertions of any such theory have to do with the relationships between rigid bodies (systems of co-ordinates), clocks, and electromagnetic processes. Insufficient consideration of this circumstance lies at the root of the difficulties which the electrodynamics of moving bodies at present encounters.

I. KINEMATICAL PART

1. Definition of Simultaneity

Let us take a system of co-ordinates in which the equations of Newtonian mechanics hold good.[3] In order to render our presentation more precise and to distinguish this system of co-ordinates verbally from others which will be introduced hereafter, we call it the "stationary system."

If a material point is at rest relatively to this system of co-ordinates, its position can be defined relatively thereto by the employment of rigid standards of measurement and the methods of Euclidean geometry, and can be expressed in Cartesian co-ordinates.

If we wish to describe the *motion* of a material point, we

give the values of its co-ordinates as functions of the time. Now we must bear carefully in mind that a mathematical description of this kind has no physical meaning unless we are quite clear as to what we understand by "time." We have to take into account that all our judgments in which time plays a part are always judgments of *simultaneous events*. If, for instance, I say, "That train arrives here at 7 o'clock," I mean something like this: "The pointing of the small hand of my watch to 7 and the arrival of the train are simultaneous events."[4]

It might appear possible to overcome all the difficulties attending the definition of "time" by substituting "the position of the small hand of my watch" for "time." And in fact such a definition is satisfactory when we are concerned with defining a time exclusively for the place where the watch is located; but it is no longer satisfactory when we have to connect in time series of events occurring at different places, or—what comes to the same thing—to evaluate the times of events occurring at places remote from the watch.

We might, of course, content ourselves with time values determined by an observer stationed together with the watch at the origin of the co-ordinates, and co-ordinating the corresponding positions of the hands with light signals, given out by every event to be timed, and reaching him through empty space. But this co-ordination has the disadvantage that it is not independent of the standpoint of the observer with the watch or clock, as we know from experience. We arrive at a much more practical determination along the following line of thought.

If at the point A of space there is a clock, an observer at A can determine the time values of events in the immediate proximity of A by finding the positions of the hands which are simultaneous with these events. If there is at the point B of space another clock in all respects resembling the one at A, it is possible for an observer at B to determine the time values of events in the immediate neighbourhood of B. But it is not possible without further assumption to compare, in respect of time, an event at A with an event at B. We have so far defined only an "A time" and a "B time." We have not defined a common "time" for A and B, for the latter cannot be defined at all unless we establish *by definition* that the "time" required by light to travel from A to B equals the "time" it requires to travel from B to A. Let a ray of light start at the

"A time" t_A from A towards B; let it at the "B time" t_B be reflected at B in the direction of A, and arrive again at A at the "A time"t'_A.

In accordance with definition the two clocks synchronize if

$$t_B - t_A = t'_A - t_B.$$

We assume that this definition of synchronism is free from contradictions, and possible for any number of points; and that the following relations are universally valid:—

1. If the clock at B synchronizes with the clock at A, the clock at A synchronizes with the clock at B.

2. If the clock at A synchronizes with the clock at B and also with the clock at C, the clocks at B and C also synchronize with each other.

Thus with the help of certain imaginary physical experiments we have settled what is to be understood by synchronous stationary clocks located at different places, and have evidently obtained a definition of "simultaneous," or "synchronous," and of "time." The "time" of an event is that which is given simultaneously with the event by a stationary clock located at the place of the event, this clock being synchronous, and indeed synchronous for all time determinations, with a specified stationary clock.

In agreement with experience we further assume the quality

$$\frac{2AB}{t'_A - t_A} = c,$$

to be a universal constant—the velocity of light in empty space.

It is essential to have time defined by means of stationary clocks in the stationary system, and the time now defined being appropriate to the stationary system, we call it "the time of the stationary system."

2. On the Relativity of Lengths and Times

The following reflexions are based on the principle of relativity and on the principle of the constancy of the velocity of light. These two principles we define as follows:—

1. The laws by which the states of physical systems undergo change are not affected, whether these changes of state be referred

to the one or the other of two systems of co-ordinates in uniform translatory motion.

2. Any ray of light moves in the "stationary" system of co-ordinates with the determined velocity *c,* whether the ray be emitted by a stationary or by a moving body. Hence

$$\text{velocity} = \frac{\text{light path}}{\text{time interval}}$$

where time interval is to be taken in the sense of the definition in § 1.

Let there be given a stationary rigid rod; and let its length be *l* as measured by a measuring-rod which is also stationary. We now imagine the axis of the rod lying along the axis of *x* of the stationary system of co-ordinates, and that a uniform motion of parallel translation with velocity *v* along the axis of *x* in the direction of increasing *x* is then imparted to the rod. We now inquire as to the length of the moving rod, and imagine its length to be ascertained by the following two operations:—

(*a*) The observer moves together with the given measuring-rod and the rod to be measured, and measures the length of the rod directly by superposing the measuring-rod, in just the same way as if all three were at rest.

(*b*) By means of stationary clocks set up in the stationary system and synchronizing in accordance with § 1, the observer ascertains at what points of the stationary system the two ends of the rod to be measured are located at a definite time. The distance between these two points, measured by the measuring-rod already employed, which in this case is at rest, is also a length which may be designated "the length of the rod."

In accordance with the principle of relativity the length to be discovered by the operation (*a*)—we will call it "the length of the rod in the moving system"—must be equal to the length *l* of the stationary rod.

The length to be discovered by the operation (*b*) we will call "the length of the (moving) rod in the stationary system." This we shall determine on the basis of our two principles, and we shall find that it differs from *l*.

Current kinematics tacitly assumes that the lengths determined by these two operations are precisely equal, or in other words, that a moving rigid body at the epoch *t* may in geometrical respects be perfectly represented by *the same* body *at rest* in a definite position.

We imagine further that at the two ends A and B of the rod, clocks are placed which synchronize with the clocks of the stationary system, that is to say that their indications correspond at any instant to the "time of the stationary system" at the places where they happen to be. These clocks are therefore "synchronous in the stationary system."

We imagine further that with each clock there is a moving observer, and that these observers apply to both clocks the criterion established in § 1 for the synchronization of two clocks. Let a ray of light depart from A at the time[5] t_A, let it be reflected at B at the time t_B, and reach A again at the time t'_A. Taking into consideration the principle of the constancy of the velocity of light we find that

$$t_B - t_A = \frac{r_{AB}}{c - v} \text{ and } t'_A - t_B = \frac{r_{AB}}{c + v}$$

where r_{AB} denotes the length of the moving rod—measured in the stationary system. Observers moving with the moving rod would thus find that the two clocks were not synchronous, while observers in the stationary system would declare the clocks to be synchronous.

So we see that we cannot attach any *absolute* signification to the concept of simultaneity, but that two events which, viewed from a system of co-ordinates, are simultaneous, can no longer be looked upon as simultaneous events when envisaged from a system which is in motion relatively to that system.

3. Theory of the Transformation of Co-ordinates and Times from a Stationary System to another System in Uniform Motion of Translation Relatively to the Former

Let us in "stationary" space take two systems of co-ordinates, i.e. two systems, each of three rigid material lines, perpendicular to one another, and issuing from a point. Let the axes of X of the two systems coincide, and their axes of Y and Z respectively be parallel. Let each system be provided with a rigid measuring-rod and a number of clocks, and let the two measuring-rods, and likewise all the clocks of the two systems, be in all respects alike.

Now to the origin of one of the two systems (k) let a constant velocity v be imparted in the direction of the increasing x of the other stationary system (K), and let this velocity be communicated to the axes of the co-ordinates, the relevant measuring-rod and the clocks. To any time of the stationary system K there then will correspond a definite position of the axes of the moving system, and from reasons of symmetry we are entitled to assume that the motion of k may be such that the axes of the moving system are at the time t (this "t" always denotes a time of the stationary system) parallel to the axes of the stationary system.

We now imagine space to be measured from the stationary system K by means of the stationary measuring-rod, and also from the moving system k by means of the measuring-rod moving with it, and that we thus obtain the co-ordinates x, y, z, and ξ, η, ζ respectively. Further, let the time t of the stationary system be determined for all points thereof at which there are clocks by means of light signals in the manner indicated in § 1; similarly let the time τ of the moving system be determined for all points of the moving system at which there are clocks at rest relatively to that system by applying the method, given in § 1, of light signals between the points at which the latter clocks are located.

To any system of values x, y, z, t, which completely defines the place and time of an event in the stationary system, there belongs a system of values ξ, η, ζ, τ, determining that event relatively to the system k, and our task is now to find the system of equations connecting these quantities.

In the first place it is clear that the equations must be *linear* on account of the properties of homogeneity which we attribute to space and time.[6]

. . . the transformation equations which have been found become

$$\tau = \beta(t - vx/c^2),$$
$$\xi = \beta(x - vt),$$
$$\eta = y,$$
$$\zeta = z,$$

where

$$\beta = 1/\sqrt{(1 - v^2/c^2)}.$$

4. Physical Meaning of the Equations Obtained in Respect to Moving Rigid Bodies and Moving Clocks

We envisage a rigid sphere[7] of radius R, at rest relatively to the moving system k, and with its centre at the origin of co-ordinates of k. The equation of the surface of this sphere moving relatively to the system K with velocity v is

$$\xi^2 + \eta^2 + \zeta^2 = R^2.$$

The equation of this surface expressed in x, y, z at the time $t = 0$ is

$$\frac{x^2}{(\sqrt{(1 - v^2/c^2)})^2} + y^2 + z^2 = R^2,$$

A rigid body which, measured in a state of rest, has the form of a sphere, therefore has in a state of motion—viewed from the stationary system—the form of an ellipsoid of revolution with the axes

$$R\sqrt{(1 - v^2/c^2)}, R, R.$$

Thus, whereas the Y and Z dimensions of the sphere (and therefore of every rigid body of no matter what form) do not appear modified by the motion, the X dimension appears shortened in the ratio $1:\sqrt{(1 - v^2/c^2)}$, i.e. the greater the value of v, the greater the shortening. For $v = c$ all moving objects—viewed from the "stationary" system—shrivel up into plain figures. For velocities greater than that of light our deliberations become meaningless; we shall, however, find in what follows, that the velocity of light in our theory plays the part, physically, of an infinitely great velocity.

It is clear that the same results hold good of bodies at rest in the "stationary" system, viewed from a system in uniform motion.

Further, we imagine one of the clocks which are qualified to mark the time t when at rest relatively to the stationary system, and the time τ when at rest relatively to the moving system, to be located at the origin of the co-ordinates of k, and so adjusted that it marks the time τ. What is the rate of this clock, when viewed from the stationary system?

Between the quantities x, t, and τ, which refer to the position of the clock, we have, evidently, $x = vt$ and

$$\tau = \frac{1}{\sqrt{(1 - v^2/c^2)}} \, (t - vx/c^2).$$

Therefore,

$$\tau = t\sqrt{(1 - v^2/c^2)} = t - (1 - \sqrt{(1 - v^2/c^2)})t$$

whence it follows that the time marked by the clock (viewed in the stationary system) is slow by $1 - \sqrt{(1 - v^2/c^2)}$ seconds per second, or—neglecting magnitudes of fourth and higher order— by $\frac{1}{2}v^2/c^2$.

From this there ensues the following peculiar consequence. If at the points A and B of K there are stationary clocks which, viewed in the stationary system, are synchronous; and if the clock at A is moved with the velocity v along the line AB to B, then on its arrival at B the two clocks no longer synchronize, but the clock moved from A to B lags behind the other which has remained at B by $\frac{1}{2}v^2/c^2$ (up to magnitudes of fourth and higher order), t being the time occupied in the journey from A to B.

It is at once apparent that this result still holds good if the clock moves from A to B in any polygonal line, and also when the points A and B coincide.

If we assume that the result proved for a polygonal line is also valid for a continuously curved line, we arrive at this result: If one of two synchronous clocks at A is moved in a closed curve with constant velocity until it returns to A, the journey lasting t seconds, then by the clock which has remained at rest the travelled clock on its arrival at A will be $\frac{1}{2}tv^2/c^2$ second slow. Thence we conclude that a balance-clock[8] at the equator must go more slowly, by a very small amount, than a precisely similar clock situated at one of the poles under otherwise identical conditions.

Relativity and the Problem of Space

ALBERT EINSTEIN, from *Relativity, the Special and the General Theory*

It is characteristic of Newtonian physics that it has to ascribe independent and real existence to space and time as well as to matter, for in Newton's law of motion the concept of acceleration appears. But in this theory, acceleration can only denote "acceleration with respect to space." Newton's space must thus be thought of as "at rest," or at least as "unaccelerated," in order that one can consider the acceleration, which appears in the law of motion as being a magnitude with any meaning. Much the same holds with time, which of course likewise enters into the concept of acceleration. Newton himself and his most critical contemporaries felt it to be disturbing that one had to ascribe physical reality both to space itself as well as to its state of motion; but there was at that time no other alternative, if one wished to ascribe to mechanics a clear meaning.

It is indeed an exacting requirement to have at all to ascribe physical reality to space, and especially to empty space. Time and again since remotest times philosophers have resisted such a presumption. Descartes argued somewhat on these lines: space is identical with extension, but extension is connected with bodies; thus there is no space without bodies and hence no empty space. The weakness of this argument lies primarily in what follows. It is certainly true that the concept of extension owes its origin to our experiences of laying out or bringing into contact solid bodies. But from this it cannot be concluded that the concept of extension may not be justified in cases which have not themselves given rise to the formation of this concept. Such

"Relativity and the Problem of Space," reprinted by permission of the publisher from *Relativity, the Special and the General Theory* by Albert Einstein, tr. by R. W. Lawson, Methuen and Company, Ltd., London, 1954.

166

an enlargement of concepts can be justified indirectly by its value for the comprehension of empirical results. The assertion that extension is confined to bodies is therefore of itself certainly unfounded. We shall see later, however, that the general theory of relativity confirms Descartes' conception in a roundabout way. What brought Descartes to his seemingly odd view was certainly the feeling that, without compelling necessity, one ought not to ascribe reality to a thing like space, which is not capable of being "directly experienced."[1]

The psychological origin of the idea of space, or of the necessity for it, is far from being so obvious as it may appear to be on the basis of our customary habit of thought. The old geometers deal with conceptual objects (straight line, point, surface), but not really with space as such, as was done later in analytical geometry. The idea of space, however, is suggested by certain primitive experiences. Suppose that a box has been constructed. Objects can be arranged in a certain way inside the box, so that it becomes full. The possibility of such arrangements is a property of the material object "box," something that is given with the box, the "space enclosed" by the box. This is something which is different for different boxes, something that is thought quite naturally as being independent of whether or not, at any moment, there are any objects at all in the box. When there are no objects in the box, its space appears to be "empty."

So far, our concept of space has been associated with the box. It turns out, however, that the storage possibilities that make up the box-space are independent of the thickness of the walls of the box. Cannot this thickness be reduced to zero, without the "space" being lost as a result? The naturalness of such a limiting process is obvious, and now there remains for our thought the space without the box, a self-evident thing, yet it appears to be so unreal if we forget the origin of this concept. One can understand that it was repugnant to Descartes to consider space as independent of material objects, a thing that might exist without matter.[2] (At the same time, this does not prevent him from treating space as a fundamental concept in his analytical geometry.) The drawing of attention to the vacuum in a mercury barometer has certainly disarmed the last of the Cartesians. But it is not to be denied that, even at this primitive stage, something unsatisfactory clings to the concept of space, or to space thought of as an independent real thing.

The ways in which bodies can be packed into space (box) are the subject of three-dimensional Euclidean geometry, whose axiomatic structure readily deceives us into forgetting that it refers to realizable situations.

If now the concept of space is formed in the manner outlined above, and following on from experience about the "filling" of the box, then this space is primarily a *bounded* space. This limitation does not appear to be essential, however, for apparently a larger box can always be introduced to enclose the smaller one. In this way space appears as something unbounded.

I shall not consider here how the concepts of the three-dimensional and the Euclidean nature of space can be traced back to relatively primitive experiences. Rather, I shall consider first of all from other points of view the rôle of the concept of space in the development of physical thought.

When a smaller box s is situated, relatively at rest, inside the hollow space of a larger box S, then the hollow space of s is a part of the hollow space of S, and the same "space," which contains both of them, belongs to each of the boxes. When s is in motion with respect to S, however, the concept is less simple. One is then inclined to think that s encloses always the same space, but a variable part of the space S. It then becomes necessary to apportion to each box its particular space, not thought of as bounded, and to assume that these two spaces are in motion with respect to each other.

Before one has become aware of this complication, space appears as an unbounded medium or container in which material objects swim around. But it must now be remembered that there is an infinite number of spaces, which are in motion with respect to each other. The concept of space as something existing objectively and independent of things belongs to prescientific thought, but not so the idea of the existence of an infinite number of spaces in motion relatively to each other. This latter idea is indeed logically unavoidable, but is far from having played a considerable rôle even in scientific thought.

But what about the psychological origin of the concept of time? This concept is undoubtedly associated with the fact of "calling to mind," as well as with the differentiation between sense experiences and the recollection of these. Of itself it is doubtful whether the differentiation between sense experience and recollection (or a mere mental image) is something psychologically directly given to us. Everyone has experienced that he has been in doubt whether he has actually experienced something with his senses or has simply dreamed about it. Probably the ability to discriminate between these alternatives first comes about as the result of an activity of the mind creating order.

An experience is associated with a "recollection," and it is considered as being "earlier" in comparison with "present experiences." This is a conceptual ordering principle for recollected experiences, and the possibility of its accomplishment gives rise to the subjective concept of time, i.e., that concept of time which refers to the arrangement of the experiences of the individual.

What do we mean by rendering objective the concept of time? Let us consider an example. A person A ("I") has the experience "it is lightning." At the same time the person A also experiences such a behavior of the person B as brings the behavior of B into relation with his own experience "it is lightning." Thus it comes about that A associates with B the experience "it is lightning." For the person A the idea arises that other persons also participate in the experience "it is lightning." "It is lightning" is now no longer interpreted as an exclusively personal experience, but as an experience of other persons (or eventually only as a "potential experience"). In this way arises the interpretation that "it is lightning," which originally entered into the consciousness as an "experience," is now also interpreted as an (objective) "event." It is just the sum total of all events that we mean when we speak of the "real external world."

We have seen that we feel ourselves impelled to ascribe a temporal arrangement to our experiences, somewhat as follows. If β is later than α and γ later than β, then γ is also later than α ("sequence of experiences"). Now what is the position in this respect with the "events" which we have associated with the experiences? At first sight it seems obvious to assume that a temporal arrangement of events exists which agrees with the temporal arrangement of the experiences. In general, and unconsciously this was done, until skeptical doubts made themselves felt.[3] In order to arrive at the idea of an objective world, an additional constructive concept still is necessary: the event is localized not only in time, but also in space.

In the previous paragraphs we have attempted to describe

how the concepts space, time, and event can be put psychologically into relation with experiences. Considered logically,
they are free creations of the human intelligence, tools of
thought, which are to serve the purpose of bringing experiences
into relation with each other, so that in this way they can be
better surveyed. The attempt to become conscious of the empirical sources of these fundamental concepts should show to
what extent we are actually bound to these concepts. In this
way we become aware of our freedom, of which, in case of
necessity, it is always a difficult matter to make sensible use.

We still have something essential to add to this sketch concerning the psychological origin of the concepts space-time-event
(we will call them more briefly "space-like," in contrast to concepts from the psychological sphere). We have linked up the
concept of space with experiences using boxes and the arrangement of material objects in them. Thus this formation of concepts already presupposes the concept of material objects (e.g.,
"boxes"). In the same way persons, who had to be introduced
for the formation of an objective concept of time, also play the
rôle of material objects in this connection. It appears to me,
therefore, that the formation of the concept of the material object must precede our concepts of time and space.

All these space-like concepts already belong to pre-scientific
thought, along with concepts like pain, goal, purpose, etc., from
the field of psychology. Now it is characteristic of thought in
physics, as of thought in natural science generally, that it endeavors in principle to make do with "space-like" concepts *alone,*
and strives to express with their aid all relations having the form
of laws. The physicist seeks to reduce colors and tones to vibrations; the physiologist, thought and pain to nerve processes, in
such a way that the physical element as such is eliminated from
the causal nexus of existence, and thus nowhere occurs as an
independent link in the causal associations. It is no doubt this
attitude, which considers the comprehension of all relations by
the exclusive use of only "space-like" concepts as being possible
in principle, that is at the present time understood by the term
"materialism" (since "matter" has lost its rôle as a fundamental
concept).

Why is it necessary to drag down from the Olympian fields of
Plato the fundamental ideas of thought in natural science, and

to attempt to reveal their earthly lineage? Answer: In order to free these ideas from the taboo attached to them, and thus to achieve greater freedom in the formation of ideas or concepts. It is to the immortal credit of D. Hume and E. Mach that they, above all others, introduced this critical conception.

Science has taken over from pre-scientific thought the concepts space, time, and material object (with the important special case "solid body"), and has modified them and rendered them more precise. Its first significant accomplishment was the development of Euclidean geometry, whose axiomatic formulation must not be allowed to blind us to its empirical origin (the possibilities of laying out or juxtaposing solid bodies). In particular, the three-dimensional nature of space as well as its Euclidean character are of empirical origin (it can be wholly filled by like constituted "cubes").

The subtlety of the concept of space was enhanced by the discovery that there exist no completely rigid bodies. All bodies are elastically deformable and alter in volume with change in temperature. The structures, whose possible configurations are to be described by Euclidean geometry, cannot therefore be characterized without reference to the content of physics. But since physics after all must make use of geometry in the establishment of its concepts, the empirical content of geometry can be stated and tested only in the framework of the whole of physics.

In this connection atomistics must also be borne in mind, and its conception of finite divisibility; for spaces of sub-atomic extension cannot be measured up. Atomistics also compels us to give up, in principle, the idea of sharply and statically defined bounding surfaces of solid bodies. Strictly speaking, there are no *precise* laws, even in the macro-region, for the possible configurations of solid bodies touching each other.

In spite of this, no one thought of giving up the concept of space, for it appeared indispensable in the eminently satisfactory whole system of natural science. Mach, in the nineteenth century, was the only one who thought seriously of an elimination of the concept of space, in that he sought to replace it by the notion of the totality of the instantaneous distances between all material points. (He made this attempt in order to arrive at a satisfactory understanding of inertia.)

The Field

In Newtonian mechanics, space and time play a dual rôle. First, they play the part of carrier or frame for things that happen in physics, in reference to which events are described by the space coordinates and the time. In principle, matter is thought of as consisting of "material points," the motions of which constitute physical happening. When matter is thought of as being continuous, this is done, as it were, provisionally in those cases where one does not wish to or cannot describe the discrete structure. In this case small parts (elements of volume) of the matter are treated similarly to material points, at least in so far as we are concerned merely with motions and not with occurrences which, at the moment, it is not possible or serves no useful purpose to attribute to motions (e.g., temperature changes, chemical processes). The second rôle of space and time was that of being an "inertial system." Inertial systems were considered to be distinguished among all conceivable systems of reference in that, with respect to them, the law of inertia claimed validity.

In this, the essential thing is that "physical reality," thought of as being independent of the subjects experiencing it, was conceived as consisting, at least in principle, of space and time on one hand, and of permanently existing material points, moving with respect to space and time, on the other. The idea of the independent existence of space and time can be expressed drastically in this way: if matter were to disappear, space and time alone would remain behind (as a kind of stage for physical happening).

This standpoint was overcome in the course of a development which, in the first place, appeared to have nothing to do with the problem of space-time, namely, the appearance of the *concept of field* and its final claim to replace, in principle, the idea of a particle (material point). In the framework of classical physics, the concept of field appeared as an auxiliary concept, in cases in which matter was treated as a continuum. For example, in the consideration of the heat conduction in a solid body, the state of the body is described by giving the temperature at every point of the body for every definite time. Mathematically, this means that the temperature T is represented as a mathematical expression (function) of the space coordinates and the time t (temperature field). The law of heat conduction is represented as a local relation (differential equation), which embraces all special cases of the conduction of heat. The temperature is here a simple example of the concept of field. This is a quantity (or a complex of quantities), which is a function of the coordinates and the time. Another example is the description of the motion of a liquid. At every point there exists at any time a velocity, which is quantitatively described by its three "components" with respect to the axes of a coordinate system (vector). The components of the velocity at a point (field components), here also are functions of the coordinates (x, y, z) and the time (t).

It is characteristic of the fields mentioned that they occur only within a ponderable mass; they serve only to describe a state of this matter. In accordance with the historical development of the field concept, where no matter was available there could also exist no field. But in the first quarter of the nineteenth century it was shown that the phenomena of the interference and the diffraction of light could be explained with astonishing accuracy when light was regarded as a wave-field, completely analogous to the mechanical vibration field in an elastic solid body. It was thus felt necessary to introduce a field that could also exist in "empty space" in the absence of ponderable matter.

This state of affairs created a paradoxical situation, because, in accordance with its origin, the field concept appeared to be restricted to the description of states in the inside of a ponderable body. This seemed to be all the more certain, inasmch as the conviction was held that every field is to be regarded as a state capable of mechanical interpretation, and this presupposed the presence of matter. One thus felt compelled, even in the space which had hitherto been regarded as empty, to assume everywhere the existence of a form of matter, which was called "ether."

The emancipation of the field concept from the assumption of its association with a mechanical carrier finds a place among the psychologically most interesting events in the development of physical thought. During the second half of the nineteenth century, in connection with the researches of Faraday and Maxwell, it became more and more clear that the description of electromagnetic processes in terms of field was vastly superior to a treatment on the basis of the mechanical concepts of material points. By the introduction of the field concept in electrodynamics, Maxwell succeeded in predicting the existence of electromagnetic waves, the essential identity of which with light waves could not be doubted, if only because of the equality of their velocity of propagation. As a result of this, optics was, in principle, absorbed by electrodynamics. *One* psychological effect of this immense success was that the field concept gradually won greater independence from the mechanistic framework of classical physics.

Nevertheless, it was at first taken for granted that electromagnetic fields had to be interpreted as states of the ether, and it was zealously sought to explain these states as mechanical ones. But as these efforts always met with frustration, science gradually became accustomed to the idea of renouncing such a mechanical interpretation. Nevertheless, the conviction still remained that electromagnetic fields must be states of the ether, and this was the position at the turn of the century.

The ether-theory brought with it the question: how does the ether behave from the mechanical point of view with respect to ponderable bodies? Does it take part in the motions of the bodies, or do its parts remain at rest relatively to each other? Many ingenious experiments were undertaken to decide this question. The following important facts should be mentioned in this connection: the "aberration" of the fixed stars in consequence of the annual motion of the earth, and the "Doppler effect," i.e., the influence of the relative motion of the fixed stars on the frequency of the light reaching us from them, for known frequencies of emission. The results of all these facts and experiments, except for one, the Michelson-Morley experiment, were explained by H. A. Lorentz on the assumption that the ether does not take part in the motions of ponderable bodies, and that the parts of the ether have no relative motions at all with respect to each other. Thus the ether appeared, as it were, as the embodiment of a space absolutely at rest. But the investigation of Lorentz accomplished still more. It explained all the electromagnetic and optical processes within ponderable bodies known at that time, on the assumption that the influence of ponderable matter on the electric field—and conversely—is due solely to the fact that the constituent particles of matter carry electrical charges, which share the motion of the particles. Concerning the experiment of

Michelson and Morley, H. A. Lorentz showed that the result obtained at least does not contradict the theory of an ether at rest.

In spite of all these beautiful successes the state of the theory was not yet wholly satisfactory, and for the following reasons. Classical mechanics, of which it could not be doubted that it holds with a close degree of approximation, teaches the equivalence of all inertial systems or inertial "spaces" for the formulation of natural laws, i.e., the invariance of natural laws with respect to the transition from one inertial system to another. Electromagnetic and optical *experiments* taught the same thing with considerable accuracy. But the foundation of electromagnetic *theory* taught that a particular inertial system must be given preference, namely, that of the luminiferous ether at rest. This view of the theoretical foundation was much too unsatisfactory. Was there no modification that, like classical mechanics, would uphold the equivalence of inertial systems (special principle of relativity)?

The answer to this question is the special theory of relativity. This takes over from the theory of Maxwell-Lorentz the assumption of the constancy of the velocity of light in empty space. In order to bring this into harmony with the equivalence of inertial systems (special principle of relativity), the idea of the absolute character of simultaneity must be given up; in addition, the Lorentz transformations for the time and the space coordinates follow for the transition from one inertial system to another. The whole content of the special theory of relativity is included in the postulate: the laws of nature are invariant with respect to the Lorentz transformations. The importance of this requirement lies in the fact that it limits the possible natural laws in a definite manner.

What is the position of the special theory of relativity in regard to the problem of space? In the first place we must guard against the opinion that the four-dimensionality of reality has been newly introduced for the first time by this theory. Even in classical physics the event is localized by four numbers, three spatial coordinates and a time coordinate; the totality of physical "events" is thus thought of as being embedded in a four-dimensional continuous manifold. But on the basis of classical mechanics this four-dimensional continuum breaks up objectively into the one-dimensional time and into three-dimensional spatial sections, the latter of which contain only simultaneous events. This resolution is the same for all inertial systems. The simultaneity of two definite events with reference to one inertial system involves the simultaneity of these events in reference to all inertial systems. This is what is meant when we say that the time of classical mechanics is absolute. According to the special theory of relativity it is otherwise. The sum total of events which are simultaneous with a selected event exist, it is true, in relation to a particular inertial system, but no longer independently of the choice of the inertial system. The four-dimensional continuum is now no longer resolvable objectively into sections, which contain all simultaneous events; "now" loses for the spatially extended world its objective meaning. It is because of this that space and time must be regarded as a four-dimensional continuum that is objectively unresolvable, if it is desired to express the purport of objective relations without unnecessary conventional arbitrariness.

Since the special theory of relativity revealed the physical equivalence of all inertial systems, it proved the untenability of the hypothesis of an ether at rest. It was therefore necessary to renounce the idea that the electromagnetic field is to be regarded as a state of a material carrier. The field thus becomes an irreducible element of physical description, irreducible in the same sense as the concept of matter in the theory of Newton.

Up to now we have directed our attention to finding in what respect the concepts of space and time were *modified* by the special theory of relativity. Let us now focus our attention on those elements which this theory has taken over from classical mechanics. Here also, natural laws claim validity only when an inertial system is taken as the basis of space-time description. The principle of inertia and the principle of the constancy of the velocity of light are valid only with respect to an *inertial system*. The field-laws also can claim to have meaning and validity only in regard to inertial systems. Thus, as in classical mechanics, space is here also an independent component in the representation of physical reality. If we imagine matter and field to be removed, inertial space or, more accurately, this space together with the associated time remains behind. The four-dimensional structure (Minkowski-space) is thought of as being the carrier of matter and of the field. Inertial spaces, with their associated times, are only privileged four-dimensional coordinate systems that are linked together by the linear Lorentz transformations. Since there exist in this four-dimensional structure no longer any sections which represent "now" objectively, the concepts of happening and becoming are indeed not completely suspended, but yet complicated. It appears therefore more natural to think of physical reality as a four-dimensional existence, instead of, as hitherto, the *evolution* of a three-dimensional existence.

This rigid four-dimensional space of the special theory of relativity is to some extent a four-dimensional analogue of H. A. Lorentz's rigid three-dimensional ether. For this theory also the following statement is valid: the description of physical states postulates space as being initially given and as existing independently. Thus even this theory does not dispel Descartes' uneasiness concerning the independent, or indeed, the *a priori* existence of "empty space." The real aim of the elementary discussion given here is to show to what extent these doubts are overcome by the general theory of relativity.

The Concept of Space in the General Theory of Relativity

This theory arose primarily from the endeavor to understand the equality of inertial and gravitational mass. We start out from an inertial system S_1, whose space is, from the physical point of view, empty. In other words, there exists in the part of space contemplated neither matter (in the usual sense) nor a field (in the sense of the special theory of relativity). With reference to S_1 let there be a second system of reference S_2 in uniform acceleration. Then S_2 is thus not an inertial system. With respect to S_2 every test mass would move with an acceleration, which is independent of its physical and chemical nature. Relative to S_2, therefore, there exists a state which, at least to a first approximation, cannot be distinguished from a gravitational field. The following concept is thus compatible with the observable facts: S_2 is also equivalent to an "inertial system"; but with respect to S_2 a (homogeneous) gravitational field is present (about the origin of which one does not worry in this connection). Thus when the gravitational field is included in the framework of the consideration, the inertial system loses its objective significance, assuming that this "principle of equivalence" can be extended to any relative motion whatsoever of the systems of reference. If it is possible to base a consistent theory on these fundamental ideas, it will satisfy of itself the fact of the equality of inertial and gravitational mass, which is strongly confirmed empirically.

Considered four-dimensionally, a non-linear transformation of the four coordinates corresponds to the transition from S_1 to S_2. The question now arises: what kind of non-linear transformations are to be permitted, or, how is the Lorentz transformation to be generalized? In order to answer this question, the following consideration is decisive.

We ascribe to the inertial system of the earlier theory this property: differences in coordinates are measured by stationary "rigid" measuring rods, and differences in time by clocks at rest. The first assumption is supplemented by another, namely, that for the relative laying out and fitting together of measuring rods at rest, the theorems on "lengths" in Euclidean geometry hold. From the results of the special theory of relativity it is then concluded, by elementary considerations, that this direct physical interpretation of the coordinates is lost for systems of reference (S_2) accelerated relatively to inertial systems (S_1). But if this is the case, the coordinates now express only the order or rank of the "contiguity" and hence also the number of dimensions of the space, but do not express any of its metrical properties. We are thus led to extend the transformations to arbitrary continuous transformations.[4] This implies the general principle of relativity: Natural laws must be covariant with respect to arbitrary continuous transformations of the coordinates. This requirement (combined with that of the greatest possible logical simplicity of the laws) limits the natural laws concerned incomparably more strongly than the special principle of relativity.

This train of ideas is based essentially on the field as an independent concept. For the conditions prevailing with respect to S_2 are interpreted as a gravitational field, without the question of the existence of masses which produce this field being raised. By virtue of this train of ideas it can also be grasped why the laws of the pure gravitational field are more directly linked with the idea of general relativity than the laws for fields of a general kind (when, for instance, an electromagnetic field is present). We have, namely, good ground for the assumption that the "field-free"

Minkowski-space represents a special case possible in natural law, in fact, the simplest conceivable special case. With respect to its metrical character, such a space is characterized by the fact that $dx_1^2 + dx_2^2 + dx_3^2$ is the square of the spatial separation, measured with a unit gauge, of two infinitesimally neighboring points of a three-dimensional "space-like" cross section (Pythagorean theorem), whereas dx_4 is the temporal separation, measured with a suitable time gauge, of two events with common (x_1, x_2, x_3). All this simply means that an objective metrical significance is attached to the quantity

$$ds^2 = dx_1^2 + dx_2^2 + dx_3^2 - dx_4^2 \qquad (1)$$

as is readily shown with the aid of the Lorentz transformations. Mathematically, this fact corresponds to the condition that ds^2 is invariant with respect to Lorentz transformations.

If now, in the sense of the general principle of relativity, this space (cf. eq. (1)) is subjected to an arbitrary continuous transformation of the coordinates, then the objectively significant quantity ds is expressed in the new system of coordinates by the relation

$$ds^2 = g_{ik} dx_i dx_k \qquad \cdot \qquad \cdot \qquad \cdot \qquad (1a)$$

which has to be summed up over the indices i and k for all combinations 11, 12, ... up to 44. The terms g_{ik} now are not constants, but functions of the coordinates, which are determined by the arbitrarily chosen transformation. Nevertheless, the terms g_{ik} are not arbitrary functions of the new coordinates, but just functions of such a kind that the form (1a) can be transformed back again into the form (1) by a continuous transformation of the four coordinates. In order that this may be possible, the functions g_{ik} must satisfy certain general covariant equations of condition, which were derived by B. Riemann more than half a century before the formulation of the general theory of relativity ("Riemann condition"). According to the principle of equivalence, (1a) describes in general covariant form a gravitational field of a special kind, when the functions g_{ik} satisfy the Riemann condition.

It follows that the law for the pure gravitational field of a general kind must be satisfied when the Riemann condition is satisfied; but it must be weaker or less restricting than the

Riemann condition. In this way the field law of pure gravitation is practically completely determined, a result which will not be justified in greater detail here.

We are now in a position to see how far the transition to the general theory of relativity modifies the concept of space. In accordance with classical mechanics and according to the special theory of relativity, space (space-time) has an existence independent of matter or field. In order to be able to describe at all that which fills up space and is dependent on the coordinates, space-time or the inertial system with its metrical properties must be thought of as existing to start with, for otherwise the description of "that which fills up space" would have no meaning.[5] On the basis of the general theory of relativity, on the other hand, space as opposed to "what fills space," which is dependent on the coordinates, has no separate existence. Thus a pure gravitational field might have been described in terms of the g_{ik} (as functions of the coordinates), by solution of the gravitational equations. If we imagine the gravitational field, i.e., the functions g_{ik}, to be removed, there does not remain a space of the type (1), but absolutely *nothing*, and also no "topological space." For the functions g_{ik} describe not only the field, but at the same time also the topological and metrical structural properties of the manifold. A space of the type (1), judged from the standpoint of the general theory of relativity, is not a space without field, but a special case of the g_{ik} field, for which—for the coordinate system used, which in itself has no objective significance—the functions g_{ik} have values that do not depend on the coordinates. There is no such thing as an empty space, i.e., a space without field. Space-time does not claim existence on its own, but only as a structural quality of the field.

Thus Descartes was not so far from the truth when he believed he must exclude the existence of an empty space. The notion indeed appears absurd, as long as physical reality is seen exclusively in ponderable bodies. It requires the idea of the field as the representative of reality, in combination with the general principle of relativity, to show the true kernel of Descartes' idea; there exists no space "empty of field."

The Principle of Equivalence

D. W. SCIAMA, from *The Unity of the Universe*

Mach bequeathed two problems to the physicists who followed him. First, he raised the issue of specifying a proper frame of reference. It was his belief that in order to insure the wide use of such a frame of reference, it would have to be described by referring to all the masses of the universe. Recently, the late physicist P. W. Bridgman challenged this thesis.

Mach's second problem arises from his analysis of Newton's bucket experiment. Mach suggested that the centrifugal force which produced the dip in the water level was produced by rotation with respect to the fixed stars. This thesis of Mach is an extremely difficult one to challenge directly. It states that the stars somehow produce centrifugal forces, but it does not specify any definite way in which this happens. Einstein and Sciama have tried to fill in this gap by offering a concrete theory of the origin of inertial forces which include centrifugal forces as a special case. Inertial forces can best be understood in relation to inertial frames. In these systems of reference, Newton's Second Law of Motion is true, to a first approximation. However, if a frame of reference accelerates with respect to an inertial frame of reference (or with respect to the fixed stars), then the force must be supplemented by so-called inertial forces in order to preserve the relation between the total force on a body and its acceleration.

Centrifugal forces are a special kind of inertial force which must be added when the frame of reference rotates with respect to the fixed stars. Einstein called attention to the fact that inertial forces are like gravitational ones in being proportional to the masses of the bodies on which they act. His Principle of Equivalence, one of the basic laws of his General Theory of Relativity, asserts that there is, or there should be, no law which is true for gravitational forces which is not also true for inertial forces (and conversely).

Mach set physicists the problem of explaining why acceleration with respect to the fixed stars produced centrifugal forces. Given Einstein's General Theory of Relativity, Mach's problem has changed. Now the problem is to explain why acceleration with respect to the fixed stars produces a gravitational type of force.

D. Sciama has explored a specific theory of gravitational force which arises from accelerating bodies, and we refer the reader to his discussion of whether the new gravitational theory can account for the presence of centrifugal and other inertial forces. The work of Sciama has a direct bearing on the question whether the Law of Inertia is true. The force of Sciama's work is to show why certain frames of reference are inertial when they do not accelerate with respect to the fixed stars. If successful, his theory would explain why the Law of Inertia holds when motions are referred to nonaccelerating frames. A fundamental law of mechanics would no longer be merely a truth—it would be an explained truth.

Some historians deny that Galileo actually dropped anything from the Leaning Tower of Pisa, but we know that he conducted a thought-experiment which convinced him that a massive body and a light body would fall with the same acceleration, and so would land together (in the absence of air resistance). This experiment involves a heavy falling body. Split this body in imagination into two halves which fall side by side. Each of these portions will have the same acceleration as a body of just half the original mass. If the halves are combined again, they will presumably continue to have this acceleration. This would mean that a body has the same acceleration as another body half as massive. Galileo generalized this result, and concluded that bodies of different mass are equally accelerated by gravity. Very accurate measurements over a wide range of masses and materials have since confirmed his reasoning.

Galileo's result has an important consequence. According to Newton's second law of motion, the acceleration of a body is equal to the ratio of the force acting on it to its mass. Since falling bodies of *different* mass have the *same* acceleration, *the gravitational force acting on them must be proportional to their mass.* This property of gravitation is in striking contrast with the action of electric and magnetic forces, which do not induce the same accelerations in all bodies. An extreme example of this is provided by neutral bodies, which they do not accelerate at all. This difference between gravitation and electromagnetism was, of course, well known to all physicists since the time of Galileo, but no one saw its significance until 1907.

Einstein realized that this similarity between gravitational and inertial forces makes it *impossible to distinguish between them.* This is best understood in terms of the way gravitational forces *can* be distinguished from electric or magnetic ones. Suppose we are told that there is a field of force present, and are asked to find out how much of it is gravitational and how much electrical. All we have to do is to measure the acceleration of one neutral body and one charged body. The acceleration of the neutral body tells us the strength of the *gravitational* part of the force. Now this part induces the same acceleration in the charged body and the neutral one. Thus the *difference* between their accelerations is a measure of the strength of the electric force

Now suppose we tried to use the same technique to distinguish between a gravitational and an inertial force. Since both these forces are proportional to the mass of the body on which they act, we would be unable to make the required distinction. Whatever body we use to measure the force, the resulting acceleration will always be the same. We can determine the total strength of the force, but we cannot tell how much of it is gravitational and how much is inertial.

Einstein was fond of illustrating this state of affairs in the following way. Consider a man enclosed in a box somewhere in space far removed from gravitational forces. Suppose that the box is suddenly pulled by a rope, so that it accelerates relative to an inertial frame. The man inside the box may choose to consider himself as at rest throughout this experiment, but then the box represents a non-inertial frame of reference. Consequently an inertial force will be acting on the box. The existence of this inertial force will be obvious to the man; if he releases an object it will accelerate away from him. The crucial point is that *this acceleration will be the same for all objects he releases,* since it is just equal and opposite to his own acceleration relative to an inertial frame. But this is exactly what would happen if, instead of being pulled by a rope, the box were acted on by a gravitational force. This means that the man will not be able to tell which of these two possibilities is the correct one!

So far this conclusion is based entirely on the similarity in the observed response of *massive* bodies to a gravitational force. The possibility clearly arises that there is some other criterion by which an inertial force can be distinguished from a gravitational one—for instance, from the behavior of light, or some subtle atomic phenomenon on a microscopic level. What Einstein did was to elevate Galileo's experiment into a principle—the famous principle of equivalence. According to this principle, *there is no criterion whatsoever by means of which an inertial force can be distinguished from a gravitational one.*

Einstein's Explanation of the Principle of Equivalence

This principle of equivalence is a plausible generalization of Galileo's discovery. At the same time it is very puzzling. Why should inertial forces mimic gravitational ones so closely that they can never be distinguished as definitely inertial? Einstein answered this question in a beautifully simple way by saying that *inertial forces are themselves gravitational in origin.* Now gravitational forces must originate from something—they must have sources in the form of lumps of matter. Which lumps of matter, then, are the sources of inertial forces? With Mach's principle in mind, Einstein had no difficulty in answering this question. Indeed, the difficulty was really the other way around—Mach had provided the sources of inertial forces but had left the nature of these forces quite obscure. All Einstein had to do was to put the two problems together. He therefore concluded that the inertial forces which arise in a non-inertial frame of reference are gravitational forces exerted by the stars.

These gravitational forces presumably cancel one another out in frames of reference at rest relative to the stars, if the stars (or rather galaxies) are disturbed more or less symmetrically. There will then be no inertial forces in such frames—that is, they must be inertial frames. On the other hand, we know that there are inertial forces acting in frames which accelerate relative to the stars. If Einstein is right, this means that the gravitational force of the stars must arise from their *acceleration* relative to the non-inertial frames. In other words, *an accelerating star must exert a different gravitational force from a non-accelerating one.*

The Origin of Inertia

D. W. SCIAMA, from *The Unity of the Universe*

In order to test Einstein's idea that inertial forces are actually gravitational forces exerted by *accelerating* stars, we need a theory which tells us how much gravitation is produced by a moving star. It was this need that led Einstein to devise his general—theory of relativity, which was published in 1915. (It should not be confused with his special theory of 1905, which has nothing to do with gravitation.) Unfortunately, although general relativity is based on simple physical principles, it is very involved mathematically—so involved, indeed, that the extent to which it incorporates Mach's principle is still a matter of controversy. . . .

Nevertheless, this need not prevent us from pursuing the problem of Mach's principle. It is possible to construct a simplified version of Einstein's theory which will still tell us how much gravitation is produced by an accelerating star, but which is sufficiently simple for its implications about inertia to be completely worked out. The full complexity of Einstein's theory is needed for some problems but fortunately not for this one.

Simple Theory of Gravitation

Our task is to construct a theory from which we can calculate the gravitational force produced by any body. The first possibility that comes to mind is simply to use Newton's law of gravitation, which asserts that the gravitational force exerted by a body decreases inversely as the square of its distance—the famous inverse square law. However, we also need to know the force exerted by a *moving* body, and this complication was not considered by Newton. He seems to have taken it for granted that the gravitational force exerted by a body does not depend on its state of motion. But we require a stationary and an accelerated body to exert different forces, if the accelerated motion of the stars is to be held responsible for inertial forces.

This is not the first occasion on which the forces exerted by stationary and moving bodies have been distinguished. In 1835, Karl Friedrich Gauss (1777–1855), a German mathematician and astronomer, discovered that a moving charge exerts a different electric force from a charge at rest. In accordance with his usual custom, he did not publish this result, and it was rediscovered in 1846 by his colleague, the German physicist Wilhelm Weber (1804–91). It has since been enshrined in Maxwell's theory of electromagnetism (1865). Of particular interest to us is the force exerted by an *accelerating* body. In the electrical case this force differs from the inverse square (Coulomb) force of a stationary charge in a characteristic way, which has the important consequence that when a charge accelerates it emits electromagnetic waves. Maxwell calculated the velocity of these waves and found it to be the same as the observed velocity of light. This result led him to his greatest discovery, that light itself consists of electromagnetic waves.

This triumph encouraged the French astronomer F. Tisserand to apply these ideas to gravitation. In 1872 he suggested that the gravitational force exerted by a moving body might obey the same laws as the electric and magnetic forces exerted by a moving charge. Tisserand did not have in mind the problem of inertia, and the only interesting result he obtained was that the planets would deviate slightly from their Newtonian orbits around the sun. Such a deviation had in fact been detected in the motion of Mercury by the French astronomer Leverrier in 1845, but Tisserand was able to reproduce only a fraction of this deviation theoretically. The motion of Mercury remained a mystery until 1915, when Einstein showed that it obeyed his new laws of gravitation.

The motion of Mercury is one of the problems which needs for its solution the full complexity of Einstein's theory. Fortunately this is not true of the problem of inertia. We shall, therefore, revive Tisserand's idea and suppose that the gravitational force exerted by a moving body obeys the same laws as the electric and magnetic forces exerted by a moving charge.

We are now in a position to calculate the gravitational force exerted by the stars when they accelerate. If our theory is sound, this force will be just equal to the inertial force postulated by Newton. Now . . . when a body is regarded as being at rest the inertial force acting on it is equal to its mass times its absolute acceleration, or rather from our present Machian point of view, its mass times the acceleration of the stars. Does Tisserand's theory of gravitation lead to this result?

To answer this question we must first determine the force exerted by *one* star; then we can add together the contributions of all the stars. Since the force exerted by a star is assumed to obey the same laws as the force exerted by a charge, we shall begin with a detailed description of electrical forces.

If a charge is stationary the force it exerts on another charge is proportional to each of the charges and inversely proportional to the square of the distance between them. This is the famous law named after Charles Coulomb, a French physicist (1736–1806), and is the analogue of Newton's inverse square law for gravitation.[1] At this stage, then, we have not gone beyond Newton's theory. It is only when we go on to consider the additional force exerted when the source charge *accelerates*, and its analogue for gravitation, that we strike new ground.

This additional force is like the Coulomb force in that it is proportional to each of the charges, but there the resemblance ends. There are two major differences and two minor ones. The first major difference is that the force decreases inversely with the distance instead of with the square of the distance—doubling the distance only halves the force instead of quartering it. The second major difference is that the force is proportional to the acceleration of the source charge (so that this force is zero when the source has no acceleration). These two new features are of fundamental importance for what follows. . . .

These two forces—the Coulomb force and the acceleration force—can be taken over directly into gravitation. The only change needed is to replace electric charge by "gravitational charge." We have learned from Newton to call it gravitational *mass*, but it can hardly be overemphasized that this expression must not be confused with inertial mass, particularly as we intend to *derive* this latter concept from our theory of gravitation. With this warning, then, we shall use the phrase "gravitational mass" to refer to the gravitational analogue of charge.

We come at last to our main task, that of calculating the total gravitational force exerted on a body by the stars when they accelerate. Now the inverse square parts of their forces cancel out by symmetry, since there is presumably the same number of stars in every direction. In contrast, the acceleration forces do not cancel out; there is a net force on the body which acts in the same direction as that of the stars' acceleration.

This net force will be proportional to the acceleration of the stars, since this is so for each individual star. This means that we have already achieved half of our task, which was to make the gravitational force of the stars equal to their acceleration times the inertial mass of the body. Our remaining task is to see whether the inertial mass of the body is equal to all the other factors besides acceleration which determine the total force of the stars.

These factors depend on the gravitational masses of the stars and on their distances. We can assume for simplicity that all the stars have the same gravitational mass, but we cannot ignore the differences in their distances. This raises a serious problem: how can we combine the forces exerted by the stars unless we know all their individual distances?

The way to approach this problem is to discover which stars make the main contribution—as a first approximation we can concentrate on them. Now we faced a similar problem earlier in this book. In order to resolve Olbers' paradox, we began by locating the stars which in his calculation made the main contribution to the light of the night sky. We found that these stars are at very great distances. The reason for this is that, although the contribution of a star decreases inversely as the square of its distance, the number of stars in a spherical layer *increases directly* as the square of the distance. Each spherical layer of stars will thus make the same contribution, except for layers at very great distances, some of whose stars lie partly behind nearer ones. This means that the main contribution comes from distant stars, since there are many more layers at large distances than at small ones. Indeed, Olbers would have obtained an infinite answer but for the shielding of very distant stars by nearer ones.

The dependence on distant sources is even more striking in our gravitational problem. In the first place, the gravitational force that arises from a star's acceleration decreases inversely as the distance rather than as the square of its distance. Since the number of stars in a spherical layer increases as the square of the distance, the contribution from a spherical layer is now *greater* for more distant layers; indeed, it is proportional to the layer's distance. Furthermore, there are now no shielding effects to keep the total from being infinite. This "gravitational Olbers' paradox" is solved in the same way as the original paradox, namely by taking into account the recession of distant sources. One can show that there is a gravitational Doppler effect which reduces the contribution from very distant regions and keeps the total finite.

. . . . .

The total gravitational force exerted by the stars depends, then, on the average gravitational density of matter at great distances, which we shall call ρ_G. It depends, in addition, on another quantity (besides their acceleration), namely the rate of expansion of the universe. This rate is relevant since it determines the size of the gravitational Doppler effect. It is measured by the value of Hubble's constant, which we shall call τ.

Finally, the force depends on the gravitational mass, m_G of the body itself. Calculation shows that the value of the force is actually $\rho_G \tau^2 m_G$ times the acceleration of the stars.[2] Now our aim was to make this force equal to the *inertial mass* (m_i) of the body times the acceleration of the stars. Thus we have derived inertial mass from our theory *if* it can be identified with the factors in our formula which multiply the acceleration. In other words, we are going to try to define inertial mass by the formula:

$$m_i = \rho_G \tau^2 m_G$$

We must now unpack the meaning of this strange-looking equation. Does the quantity m_i which is defined by it have all the properties of an inertial mass? It obviously enters the law of motion in the right way—that is, its product with acceleration gives the inertial force—indeed, this was precisely what led to our definition. In addition this definition shows that the inertial and the gravitational mass of a body (m_i and m_G) are proportional to one another. . . . this is one of the basic properties of gravitation, but whereas Newton had to *assume* it, we have been able to *derive* it.

The Gravitational Constant

We can go even further than this, for our theory also tells us the value of the constant of proportionality between m_i and m_G. To see what this implies, we shall first recall the Newtonian attitude to this quantity. According to Newton, the gravitational force between two bodies is proportional to the product of their inertial masses,[3] the constant of proportionality being known as the gravitational constant G. This constant is a measure of the *strength* of gravitation—its value determines how much gravitation is produced by a body of given inertial mass. It is generally considered to be one of the fundamental constants of nature, and it plays a vital part in our theory of Mach's principle.

The first accurate measurement of G was made by the English physicist Henry Cavendish in 1798. He succeeded in determining the force of attraction between two spheres of known inertial mass and separation. This measurement is a difficult one to make, as the force is very small. Of course gravitation seems large to us, but compare the mass of the earth with the mass of a small sphere! Some physicists have, in fact, used the attraction of the earth to determine G, but unfortunately its mass is not known very accurately. The smallness of gravitation is reflected in the smallness of the value of G: in units in which lengths are measured in centimeters and masses in grams, it has the value 6.6×10^{-8}. This means that the gravitational force between two bodies of mass one gram which are one centimeter apart accelerates them by less than one ten-millionth of a centimeter per second per second. It also means that the electrical force between two electrons is 10^{42} times greater than the gravitational force.

We are emphasizing the significance of the gravitational constant because it plays a key role in our theory of Mach's principle. Since this theory prescribes a definite value for the ratio of m_G to m_i in terms of the properties of the stars, it also determines the value of G in terms of these properties. To obtain this relation it is convenient to replace the gravitational density ρ_G of the stars by their inertial density ρ_i using, of course, the same constant of proportionality as for m_G and m_i. The final result is:

$$G\rho_i\tau^2 = 1$$

This relation is the culmination of our whole long discussion. It brings together three quantities which at first sight are completely independent of one another. Like most relations of this sort, it can be exploited in many different ways. Let us examine a few of these.

First of all, we can use the relation to calculate ρ_i, the density of matter at great distances. To do this we do not require any information obtained by telescopes, for an approximate value of Hubble's constant τ can be deduced from the amount of light in the night sky. . . . When this is combined with the Cavendish value of G we obtain an average density corresponding to about 1 hydrogen atom in every 10 liters of space. More striking than the actual value is the obtaining of a definite result at all. It means that from observations restricted to our own neighborhood we can deduce an approximate value for both Hubble's constant and the average density of matter at great distances.

· · · · ·

This result suggests that Eddington and Whittaker were probably mistaken in thinking that there is not enough matter in the universe to account for the whole of inertia. Indeed, there is a good reason why the universe should contain precisely the right amount of matter, for this amount is simply reflected in the value of G—had it contained more matter, for instance, G would have been still smaller than it actually is.

This determination of the value of G adds considerably to the fundamental significance of Mach's principle. For this principle has enabled us to obtain a *quantitative* result, namely to account for the observed value of the gravitational constant. In contrast the Newtonian theory of absolute space does not determine this constant, since it does not specify how much inertia is conferred on matter by absolute space.

Our quantitative result is also important in that it explains the *apparent* irrelevance of the properties of the stars to the inertial behavior of matter—the point which, it may be remembered, was mainly responsible for the outspoken criticisms that Mach's principle received. For we see now that the universe makes itself felt in local phenomena at just the two points where the Newtonian scheme contains arbitrary elements, namely in the choice of inertial frames and in the value of the gravitational constant.

Centrifugal and Coriolis Forces

Our theory must also be able to account for the centrifugal and Coriolis forces that act on a rotating body. If we regard such a body as at rest, then it is the stars that are rotating around it. The gravitational forces exerted by these rotating stars must be just the centrifugal and Coriolis forces. Is this requirement satis fied?

Now, according to our theory, these gravitational forces will be similar to the electric and magnetic forces exerted by a rotat-ing system of charges. The electric-type force of such a system is zero at its center but not at other points: indeed, it behaves in the required way to be a centrifugal force—that is, it acts radially away from the axis of rotation.

Furthermore, since a magnetic force acts only on a *moving* charge, its gravitational analogue acts only on a moving body. . . . Moreover, both forces act transversely to the velocity, and in opposite directions for oppositely directed velocities.[4]

We can thus identify centrifugal and Coriolis forces with the "grav-electric" and "gravo-magnetic" forces exerted by the stars when they rotate. I should emphasize that this use of the words "grav-electro" and "gravo-magnetic" is only by way of analogy, in the sense that these forces obey the same laws as electric and magnetic forces. They are in our theory completely gravitational in origin and significance.

Needless to say, we require these forces to have the right strength before they can be identified with centrifugal and Cori-olis forces. The condition that there are precisely enough stars for this to be so is just the relation that we had before, namely $G\rho\tau^2 = 1$. This helps to confirm the idea that centrifugal and Coriolis forces arise from the gravitational effects of a rotating universe.

General Conclusions

It is instructive, now, to stand a little away from the theory and to see what general conclusions can be drawn from it. Its essential feature is that it is the very distant stars which make the main contribution to inertia. In fact 80 per cent of the inertia of local matter arises from the influence of galaxies too distant to be detected by the 200-inch telescope. This means that our theoretical value for the average density of matter must relate to very distant regions of the universe. Since our value for this density is much the same as that derived from direct observation, it appears that the regions of the universe lying beyond the reach of the 200-inch telescope may not be very different from those already observed.

We have here theoretical support for Hubble's belief that in his receding galaxies he had observed a typical sample of the universe as a whole. Further support for this belief comes from the resolution of Olbers' paradox, which implies that the unsurveyed part of the universe must also be expanding, and at a rate similar to that derived by Hubble.

This shows how useful are those local phenomena which are strongly influenced by distant regions of the universe. For if we possess a theory of this influence we can use our local knowledge to discover something of the behavior of matter at great distances. Such theories, like telescopes, are tools for exploring remote regions of the universe.

The idea that distant matter can sometimes have far more influence than nearby matter may be an unfamiliar one. To make it more concrete, we give a numerical estimate of the influence of nearby objects in determining the inertia of bodies on the earth: of this inertia, the whole of the Milky Way contributes only one ten-millionth, the sun one hundred-millionth, and the earth itself one thousand-millionth!

This shows very clearly how unimportant for inertia are those parts of the universe with which we are most familiar. If this had been realized when the location of the spiral nebulae was in dispute, it could have been used to settle the dispute by showing that the Milky Way is an insignificant inhabitant of the universe.

An unfortunate consequence of this unimportance of nearby matter is that we cannot appreciably alter the inertia of a body by changing its environment. We are thus unable to use the physicist's normal experimental method of discovering influences. As a result we can easily be misled, as Newton was, into thinking that inertia is an intrinsic property of matter—one that belongs to itself alone, since it does not appear to depend in any way on its environment. This shows how difficult it will be to decide whether any apparently constant property of matter is indeed intrinsic, or whether it arises as a result of the influence of distant matter.

To see what types of influence may be at work, it is useful to have a table showing, for various possible laws of force, what is the relative importance of near and distant matter. In the following table we list in the first column the way in which each force decreases with distance. In the second column we give the fraction of the total force exerted by the stars at unit distance, on the assumption that the stars are equally spaced at this unit distance throughout the universe. Finally, in the third column, we give the extra force which would be exerted by an exceptional star, nearer than the unit distance by a factor of a million.

If we take a point at random in space, not especially near any one star, the second column estimates what fraction of the total force acting there is exerted by the nearest stars. For those special points whose distance from the nearest star is only one millionth of the unit distance, the third column gives the nearest star's additional contribution to the total force. The second column of the table shows that the switch in importance between near

Force decreases with distance as $\dfrac{1}{r^n}$	Fraction of total force which is exerted by nearest neighbors at unit distance	Extra force which is exerted by a very close neighbor at a distance of 10^{-6} of unit distance
4	$\frac{1}{2}$	10^{23}
3	$1/10$	10^{16}
2	10^{-4}	10^{7}
1	10^{-12}	10^{-8}

Total force at a typical point $= 1$

and distant matter occurs somewhere between an inverse cube and an inverse square law of force. If there is an unusually near star, the third column shows that it still dominates for an inverse square law. Since this is the law that applies in Olbers' paradox, we can understand why it is so much lighter during the day than it is at night. For the sun is a specially near star, and its light is far greater than the total amount of light from all other stars combined.

By the time we reach the inverse first power law, which characterizes our theory of inertia, even a specially near star makes a negligible contribution—only one part in a hundred million. If such a law governs other processes besides gravitation, many of the apparently intrinsic properties of matter must really arise from its interaction with the rest of the universe. Indeed, we cannot at the present time set any limits to the extent of the universe's influence on local phenomena. It is a problem for the future to discover where these limits lie.

PART II ∘∘∘

Conservation

Despite the transformations of things and the impermanence of certain structures in the world, there has from the earliest times been a more than casual interest in permanence and stability. The diurnal motions of the stars, the path of the Sun's motion among the fixed stars, even the periods of the eastward-and-westward motions of the planets relative to the Earth were noted and became the subject for much speculation. It should come as no surprise that the natural sciences have long sought for constancies; and a strong tradition in philosophy, dating at least from Plato's time, has tended to regard the real as that which is permanent, eternal, and unchanging.

The physical sciences have been extremely successful in the number and variety of constancies which they have uncovered. Let us take note of a few.

Astronomers since ancient times have noted when the planet Mars occupied diametrically opposite positions in its orbit, and have timed the planet's journey to and fro. The calculation was often made for the motion from Summer Solstice to Winter Solstice and back again. Among other things, it was assumed by the astronomers that it was Mars which had been sighted those various times. That is, a certain entity remained the same over the course of time. This may seem to be a very trivial constancy. However, no other constancies which refer to that entity can be established, unless the entity retains its identity or is preserved.

The observations on Mars not only disclosed that it took that planet different times to go from Summer to Winter Solstice and to return, but they also indicated apparent changes in its bright-

197

ness. Since two time-honored constancies were challenged by these observations, an explanation was needed. The velocity of a planet was supposed to be circular, with constant speed. Further, the planets, being considered divine and therefore unchanging, could not flame up at different times. When it was asked whether a piece of Mars would, when removed, fall toward the center of the universe or toward Mars, the answer given was that Mars could never have a piece removed, for it was unchanging. The ancient astronomers succeeded in explaining these anomalies of variation in velocity and brightness: Mars, they believed, is unchanging and does move uniformly in a circular path; but the center of that circle is not the Earth, and thus certain parts of the orbit of Mars are closer to the Earth than others, and this explains why Mars seems to be moving more rapidly and to be brighter than usual, for at those times, Mars is closer to the Earth. Aristotle's hypothesis that the Earth is at the center of the universe was no longer believed, but the permanence of Mars, its content, shape, size, and brightness were maintained.

From an early period in astronomy we can discern several kinds of constancies which have been prominent in the physical sciences. There are principles of perseverance: certain entities continue to exist throughout certain changes. There are also principles which are arithmetical or geometrical in character. They specify that, under certain conditions, the shape, size, or number of physical entities remain constant throughout time. These principles were believed to be more fundamental than any other, and attempts were made, beginning in the seventeenth century, to show how the conservation of other quantities such as mass, charge, and heat could be derived from them. For example, at one time it was believed that bodies were constituted of very fine particles which had no bodies as parts. The ultimate parts, called "atoms," were each supposed to be indestructible. Each satisfied a principle of perseverance. In addition, the number of such atoms was constant so that no new ones arose. The constancy of the number of atoms obviously goes beyond the perseverance of each. It was argued that arithmetic and geometric principles of conservation show that the total mass of bodies and even the total amount of heat remain constant. In the former case, a body was identified by the ultimate bodies which are its parts, and its mass was characterized by the total number of its atoms. This notion of mass was one of several which had wide currency between the seventeenth and the twentieth centuries. Since the total number of such atoms remained the same, it follows that the total mass was constant.

Similarly, it was believed at one time that heat was a very fine substance called "caloric," and the quantity of heat in a body was defined as the total number of its caloric atoms. Consequently, the total quantity of heat was constant. These characterizations of mass and heat later proved inadequate, so that the principles of conservation of mass and heat did not rest upon the conservation of number alone, and were in need of modification and new justifications.

There were two attempts to place the conservation of heat and charge on a sound basis: the conservation of heat was discussed in a new way with empirical support in the celebrated paper of Joule (Joule, p. 204) and the conservation of charge was given a beautiful experimental basis by Faraday in his so-called ice-pail experiment, the importance of which was later stressed by Maxwell (Faraday, p. 211; Maxwell, p. 217).

Besides the conservation of mass, heat and charge, there were principles which stated that, under certain conditions, the total linear momentum (the sum of the mass times the velocity, for each body), the total kinetic energy (the sum of half the product of the mass times the square of the velocity, for each body), and the state of motion of a body are conserved.

Descartes and Leibniz both believed that there is one physical quantity which is always conserved during motion. Descartes came close to identifying this quantity with the total linear momentum, which he thought God kept constant. Leibniz argued for the total kinetic energy or some multiple of it as the proper candidate.

The belief that there is one fundamental quantity, later called "energy," which is conserved throughout every physical change seems to be deeply rooted in our way of thinking about the world. J. R. Mayer (1814–1878) attempted to amass scientific support for the principle of the conservation of energy. But his work was regarded as obscure if not suspect, and it remained for Helmholtz to place the conservation of energy on a sure scientific footing in his epoch-making paper on the conservation of energy (1847). He showed that if the forces between bodies depend only upon some function of their mutual distance, and are directed toward each other, then a certain physical quantity would always have the same value no matter what physical processes were taking place. Faraday's reflections give us a good picture of the promise which this scientifically rooted principle held for the future of the physical sciences, and Maxwell's short letter to P. G. Tait (Maxwell, p. 245) indicates rather well the concomitant belief that energy was not only conserved but also locatable at any given time. Maxwell argued in his letter to Tait that energy had to be locatable somewhere, or else it would be physically possible to have perpetual motion, that is, an eternal source of energy which can be tapped for use, without expending much energy to do so. In other words, there would be a situation in which the total energy was not conserved.

The conservation of energy had another exciting role in physical thought. Energy was not just another physical quantity; there were various *forms of energy* not at all superficially alike. Heat and mechanical work, it was argued, are but two forms of energy. Discussion of the capacity of the steam engine, for example, seemed to show that a certain amount of heat can produce motion, but heat is not conserved. Discussions of mechanical devices seemed to show that the amount of work obtained from a machine is not equal to the amount put in—work is not conserved. According to the experiments of Joule (see Joule, p. 204), if the performance of mechanical work generates heat, then the ratio of mechanical work to heat produced is a constant. The ratio is known as the mechanical equivalent of heat. Joule also found chemical as well as electrical equivalents. The idea was suggested that these experiments involved various *forms* of energy: while the total energy in any one of its forms might not be conserved, the total energy in *all* its forms is conserved. The principle of the conservation of energy, together with some theory of the forms of energy, served to tie together many laws of conservation which, considered by themselves, had narrow scope.

There have been several ways in which the physical sciences have put order among all its conservation principles. Sometimes a group of principles are just collected together as a theory. This kind of order is obviously rudimentary. Sometimes a conservation principle can be *explained* by a theory; Newton's laws of motion do provide an explanation for the conservation of (total) linear momentum. Success of this kind can be more than partial. Helmholtz, in his celebrated paper of 1847, showed that the hypothesis of central forces together with Newton's laws of motion provided explanations of the major conservation principles of the physics of his time.

There is another way of systematizing the conservation principles, which we might call their *unification*. A theory T *unifies* two principles C_1 and C_2 if and only if, on the basis of T, it can be shown that C_1 is true if and only if C_2 is true. Einstein showed that his Special Theory of Relativity explained why energy is proportional to mass. This famous truth, $E = mc^2$ (the value of the constant c is the velocity of light *in vacuo*), is often referred to as the inertia of energy. In showing this he also showed how his theory unified the principles of conservation of energy and mass. However, neither the conservation of mass nor the conservation of energy is explained by the Special Theory of Relativity. Einstein's Special Theory unified the two principles: it showed why both principles "go together" even though it explained neither of them individually. It is obvious, though, that any theory which explains each of a group of principles also unifies them.

In addition to the explaining and the unifying of principles, there is another method of organizing them, which I shall call their "blending." We have already described how the conservation of energy in all its various forms ties together the principles related to heat and work. Although it is true that total energy is conserved when neither heat nor work are, it is also true that heat is conserved under special conditions and so, too, is mechanical work. Even if there is no theory explaining why energy is conserved, there can be a theory (thermodynamics) which, together with the energy conservation, explains why heat is conserved under its special conditions, and which also explains why mechanical work is conserved under certain conditions. This being so, one usually speaks of heat being *transformed* into mechanical work (and conversely), and both qualities come to be regarded as *forms of energy*. I shall say that the principles of heat conservation and conservation of mechanical energy have been blended.

Let us summarize these conditions. We say that a theory blends two conservation principles together if and only if (1) each of the two principles involves the physical quantities M and W, respectively; (2) there is some physical quantity Q which is a function of M and W, such that there is a well-supported principle of Q conservation; (3) the theory T together with the principle of Q conservation explains why M is conserved under special conditions, and similarly for the quantity W. We say that, in certain processes, *M is transformed into W*, and sometimes the converse, and *M and W are forms of Q*. In much the same way the reader can check that the principle of conservation of mechanical energy blends the principle of the conservation of kinetic energy—a physical quantity which is a function of the velocities of bodies (or *vis viva*)—and the conservation of a quantity called "potential energy," which is a function of the positions of bodies. There have been claims for the universal conservation of each of these quantities, but those claims are not true. What is true is that the mechanical energy, which is the *sum* of both quantities, is conserved when each fails to be. A quick check shows that the three conditions of blending hold; in fact, physicists speak of the potential energy as being transformed into kinetic energy when a body is dropped. Needless to say, both quantities are now called "forms" of energy.

When one principle blends two others, it does not in general explain either of them. But the conservation of energy, by blending principles, shows us two things It shows that heat and work are parts of a "larger" quantity which is conserved, and it explains the limits to the conservation of either quantity. The reader should not fail to read Max von Laue's beautiful study of the history of this principle.

Principles of symmetry together with principles of conservation together constituted some of the deepest truths of the physical sciences. But it was not fully realized until quite recent times that there is an intimate relation between symmetry and conservation.

Everyone would agree that some kind of *invariance* is required for there to be a symmetry. Thus, consider a circle of some definite radius, say, r. To each point of the circle, associate another point of the circle which is to the right of it. Such an association is called a clockwise rotation through an angle θ. Although each point is associated with some new point, the entire set of associated points is identical with our original set. We say in such a case that the rotation left the circle the same, or *invariant*. The circle is invariant under a rotation of θ, but the specific value θ is irrelevant in the argument above. Therefore, a circle is invariant under a whole class of transformations— the rotations about its center: the circle has radial symmetry.

A principle of symmetry states that a certain object or set of objects is invariant under a special class of transformations. Each symmetry principle is distinguished by the relevant objects and transformations. (Some writers do not speak of symmetry unless the set of transformations form a very special collection, called a "group.") Some comment is also called for by these "objects" of symmetry. It seems generally true that we speak of the symmetry of things, objects, or entities, no matter how concrete or abstract they may be. Usually, it is the symmetry of a concrete entity which is discussed: architectural monuments, snowflakes, paintings, or aerodynamic structures. Nevertheless we also speak of the symmetry of an arrangement of objects—for example, the seating of men and women around a table. Here the arrangement is certainly not to be confused with any of the objects arranged.

In a similar way, the use of symmetry in the physical sciences is not confined to concrete objects narrowly understood. Scientists refer to the symmetry of a molecule, of the intermolecular arrangements, of certain equations or other linguistic expressions, and they also study the symmetry properties of space and time. Certainly, there is no question of assimilating these into one kind of object. It would be silly, for example, to argue that all questions of symmetry in physics concern linguistic expressions. For this suggests that the investigation of symmetries is nothing more than comparative calligraphy.

The relations between conservation and symmetry in modern physics has been explored incisively only during the past four decades. Eugene Wigner has pioneered these studies. His remarks on the types of new symmetries in quantum mechanics and their relation to the fundamental properties of space and time, together with the eloquent discussion by Gerald Feinberg and Maurice Goldhaber of the limits of these symmetry and conservation principles, bring to a natural completion this set of problems central to the physical sciences.

The Mechanical Equivalent of Heat

J. P. JOULE, from *Heat and the Constitution of Elastic Fluids*[1]

According to a very influential theory of heat, now known to be false, heat was composed of a very fine substance called "caloric." This substance supposedly had a grainy structure like all matter and its ultimate parts or atoms were called caloric atoms. Although these particles repelled one another, they could combine with atoms of different gases, and they were indestructible. Finally, the quantity of heat of a body was measured by the number of caloric atoms in it, that is, by the quantity of caloric. This theory had great success in explaining all but one of the quantitative features of thermal phenomena known in the middle of the nineteenth century. According to the caloric theory of heat, the total amount of heat had to be conserved because caloric atoms were neither created nor destroyed. Despite its very sophisticated successes, this theory failed to account for one rather common phenomenon: the generation of heat by friction or rubbing. This phenomenon, together with well-attested experiences with the steam engine and other mechanical devices, convinced many physicists that neither the quantity of heat nor the quantity of mechanical work were conserved in general.

James P. Joule (1818–1889), in a series of classic experiments beginning with the famous one of 1848 reprinted below, provided what he considered sound evidence for his view that there was a mechanical equivalent of heat. In general, he believed that a certain quantity of heat could produce a certain fixed amount of mechanical work (depending upon the original amount of heat), and conversely. Stated differently, a new quantity, part of which is heat, part of which is mechanical work, remained constant. Joule spoke of the conversion of heat into work, and it was soon common enough to describe both heat and mechanical work as two forms or aspects of one and the same quantity, energy.

Reprinted from J. P. Joule, "Some Remarks on Heat and the Constitution of Elastic Fluids,"[1] *Philosophical Magazine,* Vol. XIV (4th Series), 1857.

Thus, Joule laid the sound empirical basis for a new conservation principle. He also believed that his result supported a new theory of heat, a kinetic theory of heat. According to one form of the theory which he believed in, the temperature of a gas was measured by the (rotational) motion of its atoms. He felt that his empirical researches supported a theory which related heat to mechanical work, and thus related heat to motion. And it was just such a kinetic theory which Joule was to adopt and explore further. This new theory was to blend the old principles of conservation of heat and mechanical energy (see "Introduction," Part II).

In a paper "On the Heat evolved during the Electrolysis of Water," published in the seventh volume of the Memoirs of this Society, I stated that the magneto-electrical machine enabled us to convert mechanical power into heat; and that I had little doubt that, by interposing an electro-magnetic engine in the circuit of a voltaic battery, a diminution of the quantity of heat evolved, per equivalent of chemical reaction, would be observed, and that this diminution would be proportional to the mechanical power obtained.

The results of experiments in proof of the above proposition were communicated to the British Association for the Advancement of Science, in 1843.[2] They showed that whenever a current of electricity was generated by a magneto-electrical machine, the quantity of heat evolved by that current had a constant relation to the power required to turn the machine; and, on the other hand, that whenever an engine was worked by a voltaic battery, the power developed was at the expense of the calorific power of the battery for a given consumption of zinc, the mechanical effect produced having a fixed relation to the heat lost in the voltaic circuit.

The obvious conclusion from these experiments was, that heat and mechanical power were convertible into one another; and it became therefore evident that heat is either the *vis viva* of ponderable particles, or a state of attraction or repulsion capable of generating *vis viva*.

It now became important to ascertain the mechanical equivalent of heat with as much accuracy as its importance to physical science demanded. For this purpose the magnetic apparatus was not very well adapted, and therefore I sought in the heat generated by the friction of fluids for the means of obtaining exact results. I found, first, that the expenditure of a certain amount of mechanical power in the agitation of a given fluid uniformly produced a certain fixed quantity of heat; and, secondly, that the quantity of heat evolved in the friction of fluids was entirely uninfluenced by the nature of the liquid employed; for water, oil and mercury, fluids as diverse from one another as could have been well selected, gave sensibly the same result, viz. that the quantity of heat capable of raising the temperature of a pound of water 1°, is equal to the mechanical power developed by a weight of 777 lbs. in falling through one perpendicular foot.[3]

Believing that the discovery of the equivalent of heat furnished the means of solving several interesting phænomena, I commenced, in the spring of 1844, some experiments on the changes of temperature occasioned by the rarefaction and compression of atmospheric air.[4] It had long been known that air, when forcibly compressed, evolves heat; and that, on the contrary, when air is dilated, heat is absorbed. In order to account for these facts, it was assumed that a given weight of air has a smaller capacity for heat when compressed into a small compass than when occupying a larger space. A few experiments served to show the incorrectness of this hypothesis: thus, I found that by forcing 2956 cubic inches of air, at the ordinary atmospheric pressure, into the space of $136\frac{1}{2}$ cubic inches, $13° \cdot 63$ of heat per pound of water were produced; whereas by the reverse process, of allowing the compressed air to expand from a stopcock into the atmosphere, only $4° \cdot 09$ were absorbed instead of $13° \cdot 63$, which is the quantity of heat which ought to have been absorbed, according to the generally received hypothesis. I found, also, that when strongly compressed air was allowed to escape into a vacuum, no cooling effect took place on the whole, a fact likewise at variance with the received hypothesis. On the contrary, the theory I ventured to advocate[5] was in perfect agreement with the phænomena; for the heat evolved by compressing the air was found to be the equivalent of the mechanical power employed, and *vice versa,* the heat absorbed in rarefaction was found to be the equivalent of the mechanical power developed, estimated by the weight of the column of atmospheric air displaced. In the case of compressed air expanding into a vacuum, since no mechanical power was produced, no absorption of heat was expected or found. M. Seguin has confirmed the above results in the case of steam.

The above principles lead, indeed, to a more intimate acquaintance with the true theory of the steam-engine; for they have enabled us to estimate the calorific effect of the friction of the steam in passing through the various valves and pipes, as well as that of the piston in rubbing against the sides of the cylinder; and they have also informed us that the steam, while expanding in the cylinder, loses heat in quantity exactly proportional to the mechanical force developed.[6]

The experiments on the changes of temperature produced by the rarefaction and condensation of air give likewise an insight into the constitution of elastic fluids [gases], for they show that the heat of elastic fluids is the mechanical force possessed by them; and since it is known that the temperature of a gas determines its elastic force, it follows that the elastic force, or pressure, must be the effect of the motion of the constituent particles in any gas. This motion may exist in several ways, and still account for the phænomena presented by elastic fluids. Davy, to whom belongs the signal merit of having made the first experiment absolutely demonstrative of the immateriality of heat, enunciated the beautiful hypothesis of a rotatory motion. He says, "It seems possible to account for all the phænomena of heat, if it be supposed that in solids the particles are in a constant state of vibratory motion, the particles of the hottest bodies moving with the greatest velocity and through the greatest space: that in fluids and elasic fluids, besides the vibratory motion, which must be considered greatest in the last, the particles have a motion round their own axes with different velocities, the particles of elastic fluids moving with the greatest quickness; and that in ætherial substances the particles move round their own axes, and separate from each other, penetrating in right lines through space. Temperature may be conceived to depend upon the velocity of the vibrations, increase of capacity on the motion being performed in greater space; and the diminution of temperature during the conversion of solids into fluids or gases may be explained on the idea of the loss of vibratory motion, in consequence of the revolution of particles round their axes at the moment when the body becomes fluid or aëriform, or from the loss of rapidity of vibration in consequence of the motion of the particles through greater space."[7] I have myself endeavoured to prove that a rotary motion, such as that described by Sir H. Davy, will account for the law of Boyle and Mariotte, and other phænomena presented by elastic fluids[8]; nevertheless, since the hypothesis of Herapath, in which it is assumed that the particles of a gas are constantly flying about in every direction with great velocity, the pressure of the gas being owing to the impact of the particles against any surface presented to them, is somewhat simpler, I shall employ it in the following remarks on the constitution of elastic fluids; premising, however, that the hypothesis of a rotatory motion accords equally well with the phænomena.

Let us suppose an envelope of the size and shape of a cubic foot to be filled with hydrogen gas, which, at 60° temperature and 30 inches barometrical pressure, will weigh 36·927 grs. Further, let us suppose the above quantity to be divided into three equal and indefinitely small elastic particles, each weighing 12·309 grs.; and further, that each of these particles vibrates between opposite sides of the cube, and maintains a uniform velocity except at the instant of impact; it is required to find the velocity at which each particle must move so as to produce the atmospherical pressure of 14,831,712 grs. on each of the square sides of the cube. In the first place, it is known that if a body moving with the velocity of $32\frac{1}{6}$ feet per second be opposed, during one second, by a pressure equal to its weight, its motion will be stopped, and that if the pressure be continued one second longer, the particle will acquire the velocity of $32\frac{1}{6}$ feet per second in the contrary direction. At this velocity there will be $32\frac{1}{6}$ collisions of a particle of 12·309 grs. against each side of the cubical vessel in every two seconds of time; and the pressure occasioned thereby will be $12·309 \times 32\frac{1}{6} = 395·938$ grs. Therefore, since it is manifest that the pressure will be proportional to the square of the velocity of the particles, we shall have for the velocity of the particles requisite to produce the pressure of 14,831,712 grs. on each side of the cubical vessel,

$$v = \sqrt{\left(\frac{14{,}831{,}712}{395·938}\right)32\tfrac{1}{6}} = 6225 \text{ feet per second.}$$

The above velocity will be found equal to produce the atmospheric pressure, whether the particles strike each other before they arrive at the sides of the cubical vessel, whether they strike the sides obliquely, and thirdly, into whatever number of particles the 36·927 grs. of hydrogen are divided.

If only one-half the weight of hydrogen, or 18·4635 grs., be enclosed in the cubical vessel, and the velocity of the particles be as before, 6225 feet per second, the pressure will manifestly be only one-half of what it was previously, which shows that the law of Boyle and Mariotte flows naturally from the hypothesis.

The velocity above named is that of hydrogen at the temperature of 60°; but we know that the pressure of an elastic fluid at 60° is to that at 32° as 519 is to 491. Therefore the velocity of the particles at 60° will be to that at 32° as $\sqrt{519} : \sqrt{491}$, which shows that the velocity at the freezing temperature of water is 6055 feet per second.

In the above calculations it is supposed that the particles of hydrogen have no sensible magnitude, otherwise the velocity corresponding to the same pressure would be lessened.

Since the pressure of a gas increases with its temperature in arithmetical progression, and since the pressure is proportional to the square of the velocity of the particles, in other words, to their *vis viva*, it follows that the absolute temperature, pressure, and *vis viva* are proportional to one another, and that the zero of temperature is 491° below the freezing-point of water. Further, the absolute heat of the gas, or, in other words, its capacity, will be represented by the whole amount of *vis viva* at a given temperature. The specific heat may therefore be determined in the following simple manner:—

The velocity of the particles of hydrogen, at the temperature of 60°, has been stated to be 6225 feet per second, a velocity equivalent to a fall from the perpendicular height of 602,342 feet. The velocity of 61° will be $6225 \sqrt{\dfrac{520}{519}} = 6230 \cdot 93$ feet per second, which is equivalent to a fall of 603,502 feet. The difference between the above falls is 1160 feet, which is therefore the space through which 1 lb. of pressure must operate upon each pound of hydrogen, in order to elevate its temperature one degree. But our mechanical equivalent of heat shows that 770 feet is the altitude representing the force required to raise the temperature of water one degree; consequently the specific heat of hydrogen will be $\dfrac{1160}{770} = 1 \cdot 506$, calling that of water unity.

The specific heats of the gases will be easily deduced from that of hydrogen; for the whole *vis viva* and capacity of equal bulks of the various gases will be equal to one another; and the velocity of the particles will be inversely as the square root of the specific gravity. Hence the specific heat will be inversely proportional to the specific gravity, a law which has been arrived at experimentally by De la Rive and Marcet.

In the following Table I have placed the specific heats of various gases determined in the above manner, in juxtaposition with the experimental results of Delaroche and Berard reduced to constant volume:—

	Experimental specific heat.	Theoretical specific heat.
Hydrogen . . .	2·352	1·506
Oxygen	0·168	0·094
Nitrogen . . .	0·195	0·107
Carbonic oxide . .	0·158	0·068

The experimental results of Delaroche and Berard are invariably higher than those demanded by the hypothesis. But it must be observed, that the experiments of Delaroche and Berard, though considered the best that have hitherto been made, differ considerably from those of other philosophers. I believe, however, that the investigation undertaken by M. V. Regnault, for the French Government, will embrace the important subject of the capacity of bodies for heat, and that we may shortly expect a new series of determinations of the specific heat of gases, characterized by all the accuracy for which that distinguished philosopher is so justly famous. Till then, perhaps, it will be better to delay any further modifications of the dynamical theory, by which its deductions may be made to correspond more closely with the results of experiment[9].

Conservation of Charge

MICHAEL FARADAY, from *On Static Electrical Inductive Action*

Michael Faraday (1791–1867) seems to have been the second after
Newton to introduce a famous bucket experiment into science.
In his celebrated 1843 paper, he described the so-called series of
ice-pail experiments in which he tried to give sound empirical support for the conservation of a physical quantity—the charge on a
body. There were empirical procedures for determining whether a
body was charged, and even a measure of the magnitude of
charge, using the degree of divergence of the two gold leaves of a
gold leaf electroscope. Conservation required that the charge
on an isolated or insulated body remain constant throughout time.
Also, the total charge of an insulated system of charged bodies
should remain the same, no matter how the bodies interact with
one another. Faraday tried to demonstrate the conservation of
charge, whether the charge was produced by induction, friction,
or any other of the then known ways of producing charged bodies.
The simplicity of these experiments and the elegance and force-
fulness of their sequence make them classic. Some thirty years later,
J. C. Maxwell was to describe and comment upon Faraday's experiments, and draw out some of the implications of that work
for a theory of electrodynamics (see Maxwell, p. 217).

To R. Phillips, Esq., F.R.S.

DEAR PHILLIPS,

Perhaps you may think the following experiments worth notice;
their value consists in their power to give a very precise and decided idea to the mind respecting certain principles of inductive
electrical action, which I find are by many accepted with a degree
of doubt or obscurity that takes away much of their importance:
they are the expression and proof of certain parts of my view of

Reprinted from *Philosophical Magazine*, 1843.

211

induction.[1] Let A in the diagram [Fig. 1] represent an insulated
pewter ice-pail ten and a half inches high and seven inches diameter, connected by a wire with a delicate gold-leaf electrometer E,
and let C be a round brass ball insulated by a dry thread of white

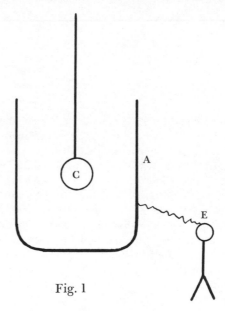

Fig. 1

silk, three or four feet in length, so as to remove the influence
of the hand holding it from the ice-pail below. Let A be perfectly
discharged, then let C be charged at a distance by a machine or
Leyden jar, and introduced into A as in the figure. If C be positive, [the gold leaves of] E also will diverge positively; if C be
taken away, E will collapse perfectly, the apparatus being in good
order. As C enters the vessel A the divergence of E will increase
until C is about three inches below the edge of the vessel, and will
remain quite steady and unchanged for any lower distance. This
shows that at that distance the inductive action of C is entirely
exerted upon the interior of A, and not in any degree directly
upon external objects. If C be made to touch the bottom of A, *all*
its charge is communicated to A; there is no longer any inductive
action between C and A, and C, upon being withdrawn and examined, is found perfectly discharged.

These are all well-known and recognized actions, but being a
little varied, the following conclusions may be drawn from them.

If C be merely suspended in A, it acts upon it by induction, evolving electricity of its own kind on the outside of A; but if C touch A its electricity is then communicated to it, and the electricity that is afterwards upon the outside of A may be considered as that which was originally upon the carrier C. As this change, however, produces no effect upon the leaves of the electrometer, it proves that the electricity *induced* by C and the electricity *in* C are accurately equal in amount and power.

Again, if C charged be held equidistant from the bottom and sides of A at one moment, and at another be held as close to the bottom as possible without discharging to A, still the divergence remains absolutely unchanged, showing that whether C acts at a considerable distance or at the very smallest distance, the amount of its force is the same. So also if it be held excentric and near to the side of the ice-pail in one place, so as to make the inductive action take place in lines expressing almost every degree of force in different directions, still the sum of their forces is the same constant quantity as that obtained before; for the leaves alter not. Nothing like expansion or coercion of the electric force appears under these varying circumstances.

I can now describe experiments with many concentric metallic vessels arranged as in the diagram [Fig. 2], where four ice-pails

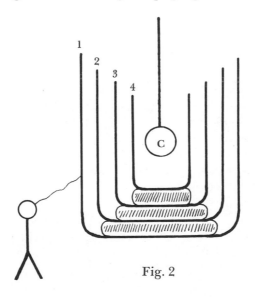

Fig. 2

are represented insulated from each other by plates of shell-lac on which they respectively stand. With this system the charged carrier C acts precisely as with the single vessel, so that the intervention of many conducting plates causes no difference in the amount of inductive effect. If C touch the inside of vessel 4, still the leaves are unchanged. If 4 be taken out by a silk thread, the leaves perfectly collapse; if it be introduced again, they open out to the same degree as before. If 4 and 3 be connected by a wire let down between them by a silk thread, the leaves remain the same, and so they still remain if 3 and 2 be connected by a similar wire; yet all the electricity originally on the carrier and acting at a considerable distance, is now on the outside of 2, and acting through only a small non-conducting space. If at last it be communicated to the outside of 1, still the leaves remain unchanged.

Again, consider the charged carrier C in the centre of the system, the divergence of the electrometer measures its inductive influence; this divergence remains the same whether 1 be there alone, or whether all four vessels be there; whether these vessels be separate as to insulation, or whether 2, 3 and 4 be connected so as to represent a very thick metallic vessel, or whether all four vessels be connected.

Again, if in place of the metallic vessels 2, 3, 4, a thick vessel of shell-lac or of sulphur be introduced, or if any other variation in the character of the substance within the vessel 1 be made, still not the slightest change is by that caused upon the divergence of the leaves.

If in place of one carrier many carriers in different positions are within the inner vessel, there is no interference of one with the other; they act with the same amount of force outwardly as if the electricity were spread uniformly over one carrier, however much the distribution on each carrier may be disturbed by its neighbours. If the charge of one carrier be by contact given to vessel 4 and distributed over it, still the others act through and across it with the same final amount of force; and no state of charge given to any of the vessels 1, 2, 3, or 4, presents a charged carrier introduced within 4 acting with precisely the same amount of force as if they were uncharged. If pieces of shell-lac, slung by white silk thread and excited, be introduced into the vessel, they act exactly as the metallic carriers, except that their charge cannot be communicated by contact to the metallic vessels.

Thus a certain amount of electricity acting within the centre of the vessel A exerts exactly the same power externally, whether it act by induction through the space between it and A, or whether it be transferred by conduction to A, so as absolutely to destroy the previous induction within. Also, as to the inductive action, whether the space between C and A be filled with air, or with shell-lac or sulphur, having above twice the specific inductive capacity of air; or contain many concentric shells of conducting matter; or be nine-tenths filled with conducting matter, or be metal on one side and shell-lac on the other; or whatever other means be taken to vary the forces, either by variation of distance or substance, or actual charge of the matter in this space, still the amount of action is precisely the same.

Hence if a body be charged, whether it be a particle or a mass, there is nothing about its action which can at all consist with the idea of exaltation or extinction; the amount of force is perfectly definite and unchangeable: or to those who in their minds represent the idea of the electric force by a fluid, there ought to be no notion of the compression or condensation of this fluid within itself, or of its coercibility, as some understand that phrase. The only mode of affecting this force is by connecting it with force of the same kind, either in the same or the contrary direction. If we oppose to it force of the contrary kind, we may *by discharge* neutralize the original force, or we may *without discharge* connect them by the simple laws and principles of static induction; but away from induction, which is *always of the same kind,* there is no other state of the power in a charged body; that is, there is no state of static electric force corresponding to the terms of *simulated* or *disguised* or *latent* electricity away from the ordinary principles of inductive action; nor is there any case where the electricity is *more latent* or *more disguised* than when it exists upon the charged conductor of an electrical machine and is ready to give a powerful spark to any body brought near it.

A curious consideration arises from this perfection of inductive action. Suppose a thin uncharged metallic globe two or three feet in diameter, insulated in the middle of a chamber, and then suppose the space within this globe occupied by myriads of little vesicles or particles charged alike with electricity (or differently), but each insulated from its neighbour and the globe; their inductive power would be such that the outside of the globe would be charged with a force equal to the sum of *all* their forces, and any part of this globe (not charged of itself) would give as long and powerful a spark to a body brought near it as if the electricity of all the particles near and distant were on the surface of the globe itself. If we pass from this consideration to the case of a cloud, then, though we cannot altogether compare the external surface of the cloud to the metallic surface of the globe, yet the previous inductive effects upon the *earth* and its buildings are the same; and when a charged cloud is over the earth, although its electricity may be diffused over every one of its particles, and no important part of the *inductric* charge be accumulated upon its under surface, yet the induction upon the earth will be as strong as if all that portion of force which is directed towards the earth *were* upon that surface; and the state of the earth and its tendency to discharge to the cloud will also be as strong in the former as in the latter case. As to whether lightning-discharge begins first at the cloud or at the earth, that is a matter far more difficult to decide than is usually supposed; theoretical notions would lead me to expect that in most cases, perhaps in all, it begins at the earth. I am,

My dear Phillips, ever yours,

M. FARADAY.

Royal Institution,
4th Feb. 1843.

Conservation and the Meaning of Charge

J. C. MAXWELL, from *A Treatise on Electricity and Magnetism*

Tradition has it that Faraday, brilliant experimental physicist that he was, was no master of mathematical analysis. J. C. Maxwell (1831–1879), as the story goes, saw in some of Faraday's experimental results relations to which he gave mathematical expression. These results, then, in mathematical form became part of Maxwell's theory of electrodynamic phenomena. And the theory in turn had many well-confirmed deductions, one of which was the conservation of charge.

Maxwell's discussion of the ice-bucket experiments makes it clear that the conservation of charge (or "electrification," as Maxwell referred to it) has been well supported by Faraday's experiments. But there are further problems raised by this result.

According to Maxwell, the charge on a body, whether positive or negative, was shown to be a genuine physical quantity, capable of measurement and subject to mathematical (algebraic) analysis. But the discovery of a law of conservation does not terminate physical inquiry. Maxwell raised a new series of questions which presupposed that the charge was a conserved quantity. What kind of a physical quantity is charge, he asked?

Some physicists thought of quantity geometrically. That is, they assumed that there were one or more special substances, and a physical quantity was the amount of the relevant substance(s). Thus mass was thought of as the volume of matter in a body. Heat, when it was believed to be a physical quantity, was also supposed to be a certain caloric substance (the temperature was its density). Electricity too, was regarded as substantial. Other physicists thought of physical quantities along arithmetic rather than geometrical lines. Thus there were those who thought that

Reprinted from J. C. Maxwell: *A Treatise on Electricity and Magnetism*, Vol. I, 1904, by permission of the Clarendon Press, Oxford.

the mass of a body was simply the number of certain atoms within it, each of those atoms contributing one unit of mass to the total. It was later to turn out that all charges were multiples of a fundamental unit of charge; this kind of result would certainly have lent support to an arithmetical interpretation of charge.

Maxwell entertained the question of whether charge was a substance (or two), or some arithmetical kind of quantity, or a form of energy (like heat) (see Joule, p. 204), or energy, or perhaps a very different kind of physical quantity altogether. His question has sometimes been asked in the form "What is the nature of electricity?" but one should not be misled by this form of the question. It appears like a request for a definition of "electricity," but Maxwell intended this question to be one which invited further experimental and theoretical inquiry. His own answer was a negative one. At least he was sure that charge was more like mass than like heat or energy. For heat was a form of energy which, unlike charge, failed to satisfy the equation of continuity. This fact was of importance for Maxwell. If a certain quantity of charge is contained within a closed surface and there is either an increase or decrease in the amount of charge within the surface, then the increase or decrease in charge can be traced passing through the boundary into space. Mass also satisfied the equation of continuity. But heat did not; the quantity of heat within a closed surface could shift in amount, without the passage of a corresponding locatable increase or decrease of heat out of the enclosed area. As for energy, Maxwell had his doubts whether that quantity was conserved, so that likening charge to energy was a dubious step (see, however, Maxwell, p. 245). Maxwell's question seems to be this: Now that we know that charge is conserved, how can we explain that fact?

The first edition of the *Treatise* appeared in 1873.

Electrification by Induction[1]

28.] EXPERIMENT II.[2]   Let a hollow vessel of metal be hung up by white silk threads, and let a similar thread be attached to the lid of the vessel so that the vessel may be opened or closed without touching it. [See Fig. 1.]

Let the pieces of glass and resin be similarly suspended and electrified as before.

Let the vessel be originally unelectrified, then if an electrified piece of glass is hung up within it by its thread without touching

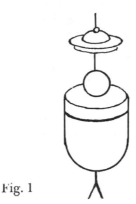

Fig. 1

the vessel, and the lid closed, the outside of the vessel will be found to be vitreously electrified, and it may be shewn that the electrification outside of the vessel is exactly the same in whatever part of the interior space the glass is suspended.[3]

If the glass is now taken out of the vessel without touching it, the electrification of the glass will be the same as before it was put in, and that of the vessel will have disappeared.

This electrification of the vessel, which depends on the glass being within it, and which vanishes when the glass is removed, is called electrification by Induction.

Similar effects would be produced if the glass were suspended near the vessel on the outside, but in that case we should find an electrification, vitreous in one part of the outside of the vessel and resinous in another. When the glass is inside the vessel the whole of the outside is vitreously and the whole of the inside resinously electrified.

Electrification by Conduction

29.] EXPERIMENT III. Let the metal vessel be electrified by induction, as in the last experiment, let a second metallic body be suspended by white silk threads near it, and let a metal wire, similarly suspended, be brought so as to touch simultaneously the electrified vessel and the second body.

The second body will now be found to be vitreously electrified, and the vitreous electrification of the vessel will have diminished.

The electrical condition has been transferred from the vessel to the second body by means of the wire. The wire is called a *conductor* of electricity, and the second body is said to be *electrified by conduction*.

Conductors and Insulators

EXPERIMENT IV.   If a glass rod, a stick of resin or gutta-percha, or a white silk thread, had been used instead of the metal wire, no transfer of electricity would have taken place. Hence these latter substances are called Non-conductors of electricity. Non-conductors are used in electrical experiments to support electrified bodies without carrying off their electricity. They are then called Insulators.

The metals are good conductors; air, glass, resins, gutta-percha, vulcanite, paraffin, &c. are good insulators; but, as we shall see afterwards, all substances resist the passage of electricity, and all substances allow it to pass, though in exceedingly different degrees. This subject will be considered when we come to treat of the motion of electricity. For the present we shall consider only two classes of bodies, good conductors, and good insulators.

In Experiment II an electrified body produced electrification in the metal vessel while separated from it by air, a non-conducting medium. Such a medium, considered as transmitting these electrical effects without conduction, has been called by Faraday a Dielectric medium, and the action which takes place through it is called Induction.

In Experiment III the electrified vessel produced electrification in the second metallic body through the medium of the wire. Let us suppose the wire removed, and the electrified piece of glass taken out of the vessel without touching it, and removed to a sufficient distance. The second body will still exhibit vitreous electrification, but the vessel, when the glass is removed, will have resinous electrification. If we now bring the wire into contact with both bodies, conduction will take place along the wire, and all electrification will disappear from both bodies, shewing that the electrification of the two bodies was equal and opposite.

30.] EXPERIMENT V.   In Experiment II it was shown that if a piece of glass, electrified by rubbing it with resin, is hung up in an insulated metal vessel, the electrification observed outside does not depend on the position of the glass. if we now introduce the piece of resin with which the glass was rubbed into the same vessel, without touching it or the vessel, it will be found that there is no electrification outside the vessel. From this we conclude that the electrification of the resin is exactly equal and opposite to that of the glass. By putting in any number of bodies, electrified in any way, it may be shewn that the electrification of the outside of the vessel is that due to the algebraic sum of all the electrifications, those being reckoned negative which are resinous. We have thus a practical method of adding the electrical effects of several bodies without altering their electrification.

31.] EXPERIMENT VI. Let a second insulated metallic vessel, B, be provided, and let the electrified piece of glass be put into the first vessel A, and the electrified piece of resin into the second vessel B. Let the two vessels be then put in communication by the metal wire, as in Experiment III. All signs of electrification will disappear.

Next, let the wire be removed, and let the pieces of glass and of resin be taken out of the vessels without touching them. It will be found that A is electrified resinously and B vitreously.

If now the glass and the vessel A be introduced together into a larger insulated metal vessel C, it will be found that there is no electrification outside C. This shews that the electrification of A is exactly equal and opposite to that of the piece of glass, and that of B may be shewn in the same way to be equal and opposite to that of the piece of resin.

We have thus obtained a method of charging a vessel with a quantity of electricity exactly equal and opposite to that of an electrified body without altering the electrification of the latter, and we may in this way charge any number of vessels with exactly equal quantities of electricity of either kind, which we may take for provisional units.

32.] EXPERIMENT VII. Let the vessel B, charged with a quantity of positive electricity, which we shall call, for the present, unity, be introduced into the larger insulated vessel C without touching it. It will produce a positive electrification on the outside of C. Now let B be made to touch the inside of C. No change of the external electrification will be observed. If B is now taken out of C without touching it, and removed to a sufficient distance, it will be found that B is completely discharged, and that C has become charged with a unit of positive electricity.

We have thus a method of transferring the charge of B to C.

Let B be now recharged with a unit of electricity, introduced into C already charged, made to touch the inside of C, and removed. It will be found that B is again completely discharged, so that the charge of C is doubled.

If this process is repeated, it will be found that however highly C is previously charged, and in whatever way B is charged, when B is first entirely enclosed in C, then made to touch C, and finally removed without touching C, the charge of B is completely transferred to C, and B is entirely free from electrification.

This experiment indicates a method of charging a body with any number of units of electricity. We shall find, when we come to the mathematical theory of electricity, that the result of this experiment affords an accurate test of the truth of the theory.[4]

33.] Before we proceed to the investigation of the law of electrical force, let us enumerate the facts we have already established.

By placing any electrified system inside an insulated hollow conducting vessel, and examining the resultant effect on the outside of the vessel, we ascertain the character of the total electrification of the system placed inside, without any communication of electricity between the different bodies of the system.

The electrification of the outside of the vessel may be tested with great delicacy by putting it in communication with an electroscope.

We may suppose the electroscope to consist of a strip of gold leaf hanging between two bodies charged, one positively, and the other negatively. If the gold leaf becomes electrified it will incline towards the body whose electrification is opposite to its own. By increasing the electrification of the two bodies and the delicacy of the suspension, an exceedingly small electrification of the gold leaf may be detected.

When we come to describe electrometers and multipliers we shall find that there are still more delicate methods of detecting electrification and of testing the accuracy of our theories, but at present we shall suppose the testing to be made by connecting the hollow vessel with a gold leaf electroscope.

This method was used by Faraday in his very admirable demonstration of the laws of electrical phenomena.[5]

34.] I. The total electrification of a body, or system of bodies, remains always the same, except in so far as it receives electrification from or gives electrification to other bodies.

In all electrical experiments the electrification of bodies is found to change, but it is always found that this change is due to want of perfect insulation, and that as the means of insulation are improved, the loss of electrification becomes less. We may therefore assert that the electrification of a body placed in a perfectly insulating medium would remain perfectly constant.

II. When one body electrifies another by conduction, the total electrification of the two bodies remains the same, that is, the one loses as much positive or gains as much negative electrification as the other gains of positive or loses of negative electrification.

For if the two bodies are enclosed in the hollow vessel, no change of the total electrification is observed.

III. When electrification is produced by friction, or by any other known method, equal quantities of positive and negative electrification are produced.

For the electrification of the whole system may be tested in the hollow vessel, or the process of electrification may be carried on within the vessel itself, and however intense the electrification of the parts of the system may be, the electrification of the whole, as indicated by the gold leaf electroscope, is invariably zero.

The electrification of a body is therefore a physical quantity capable of measurement, and two or more electrifications can be combined experimentally with a result of the same kind as when two quantities are added algebraically. We therefore are entitled to use language fitted to deal with electrification as a quantity as well as a quality, and to speak of any electrified body as 'charged with a certain quantity or positive or negative electricity.'

35.] While admitting electricity, as we have now done, to the rank of a physical quantity, we must not too hastily assume that it is, or is not, a substance, or that it is, or is not, a form of energy, or that it belongs to any known category of physical quantities. All that we have hitherto proved is that it cannot be created or annihilated, so that if the total quantity of electricity within a closed surface is increased or diminished, the increase or diminution must have passed in or out through the closed surface.

This is true of matter, and is expressed by the equation known as the Equation of Continuity in Hydrodynamics.

It is not true of heat, for heat may be increased or diminished within a closed surface, without passing in or out through the surface, by the transformation of some other form of energy into heat, or of heat into some other form of energy.

It is not true even of energy in general if we admit the immediate action of bodies at a distance. For a body outside the closed surface may make an exchange of energy with a body within the surface. But if all apparent action at a distance is the result of the action between the parts of an intervening medium, it is conceivable that in all cases of the increase or diminution of the energy within a closed surface we may be able, when the nature of this action of the parts of the medium is clearly understood, to trace the passage of the energy in or out through that surface.[6]

Conservation of Motion

LEIBNIZ, from *The Quantity of Motion*

Both Gottfried Leibniz and René Descartes agreed that despite all
the changes which the totality of bodies undergo, there is some
physical quantity which remains constant. Obviously, such a quantity would be one of the key concepts of any explanatory and
descriptive scheme for the motion of bodies. The importance of
identifying this conserved quantity was evident.

There were two candidates: motive force and quantity of motion. Descartes had claimed that the total quantity of motion is
constant, where the quantity of motion of a body is the sum of
the magnitude of the body multiplied by its speed. Leibniz, on
the other hand, argued that in some cases such as the action of
simple machines like pulleys, levers, and the like, the motive force
is identical with the quantity of motion. But there are cases of
motion where the two quantities are not the same. The case he
had in mind was that of one unit of mass falling freely through 4
feet, compared with a body of four units of mass falling freely
through a height of 1 foot. Everyone agrees, he suggested, on a
number of propositions: first, that in free fall a body acquires a
force which is equal to the force sufficient to raise it to the height
from which it fell; second, that the force required to raise a unit
mass 4 feet high is the same as that which can raise a body of four
units of mass 1 foot high.

From these two statements it follows that the motive force acquired by each of these bodies in falling must be the same. However, the quantity of motion of these bodies is not the same—the
use of Galileo's Law of Falling Bodies shows that the body which
fell 4 feet had twice the velocity and therefore half the quantity
of motion of the other body. If there were a device so that one
body's falling catapulted the other vertically, then the motive force

"Brevis Demonstratic Erroris," *Acta Eruditorum,* 1686. Reprinted by permission of the publishers from William Francis Magie, *A Source Book in
Physics,* Cambridge, Mass.: Harvard University Press, Copyright 1935, 1963, by
the President and Fellows of Harvard College.

would be constant but not the quantity of motion. Finally, Leibniz suggested that since the motive force of a body is equal to the force needed to raise the body a certain height, the motive force of a body h units above ground ought to be measured by the product of the height of the body and its mass. The reader can easily verify for himself that this product is the same for the two falling bodies.

Since we are concerned with free fall, it is worth noting that, according to Galileo, the height a body falls at any time is proportional to the square of its velocity (at that time). Instead of the product of a body's magnitude and height above the Earth, one could also use the product of the body's magnitude and the square of its velocity. This quantity was known as the *vis viva* of the body. Again, the reader can verify that the *vis viva* for the two falling bodies is the same. The *vis viva* is proportional to what is now referred to as the kinetic energy of a body. Theories of mechanics which have tried to use energy as one of their fundamental concepts have justly referred to Leibniz as the first to conceive of a science of motion which rested on such a basis.

Quantity of Motion

A short demonstration of a remarkable error made by Descartes and others in that they affirm it to be a law of nature that always the same quantity of motion is conserved by God; which law they make improper use of in applying it to mechanics.

Most mathematicians, when they see, in the cases of the five mechanical powers, that velocity and mass are mutually compensated, generally estimate the motive force by the quantity of motion or by the product of the mass of the body into its velocity. Or to speak more mathematically, the forces of two bodies (of the same sort) which are set in motion, and which act both by reason of their masses and their motions, they say, are in a ratio compounded of the bodies or masses and of the velocities which they possess. And so it may be agreeable to reason that the same totality of motive power is conserved in nature: and is neither diminished, since we see that no force is lost by a body, but is transferred to some other body; nor increased, because surely perpetual mechanical motion never occurs and no machine or even the world is able to maintain its force without a new external impulse; whence it happens that Descartes, who considered *motive force* and *quantity of motion* as equivalent, affirmed that the same quantity of motion was always conserved by God in the world.

But I, that I may show how much difference there is between these two ideas, assume, *first*, that a body falling from a certain height acquires a force sufficient to raise it to the same height, if it is given the proper direction and no external forces interfere: for example, that a pendulum will return precisely to the height from which it has been released, unless the resistance of the air and other slight obstacles absorb some of its strength, which we need not consider. I assume, *secondly*, that as much force is needed to raise a body A weighing one pound to the height CD of four ells, as to raise a body B weighing four pounds to the height EF of one ell. These assumptions are conceded by the Cartesians as well as by other philosophers and mathematicians of our times. Hence it follows that the body A [Fig 1] let fall

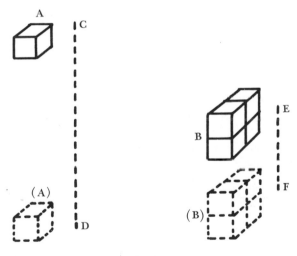

Fig. 1

from the height CD acquires exactly as much force as the body B let fall from the height EF. For the body A, after that by its fall from C it reaches D, there has the force of ascending again to C, by assumption 1, that is, the force sufficient to raise a body weighing one pound (that is, its own body) to the height of four ells. And similarly the body B, after that by its fall E it reaches F, there has the force of ascending again to E by assumption 1, that is, the force sufficient to raise a body weighing four pounds (that is its own body) to the height of one ell. Therefore by assumption 2 the force of the body A when it reaches D and the force of the body B when it reaches F are equal.

Now let us see if the quantity of motion also is the same for both. And here quite unexpectedly a great difference appears. This I show as follows: It has been demonstrated by Galileo that the velocity acquired by the fall CD is twice the velocity acquired by the fall EF. If therefore we multiply the body A, which may be taken as one, by its velocity, which may be taken as two, the product or the quantity of motion will be two. Again if we multiply the body B, which may be taken as four, by its velocity, which may be taken as one, the product or the quantity of motion will be four. Therefore the quantity of motion of the body A when at the point D is half that of the body B when at the point F, and yet just now the forces of both of these bodies have been found to be equal. And so there is a great difference between motive force and quantity of motion, so that one of these magnitudes cannot be determined from the other; which we have undertaken to show. From this it appears in what way the force should be estimated, from the quantity of the effect which it is able to produce; for example, from the height to which it can lift a heavy body of known magnitude and nature, not from the velocity which it can impress on the body. For there is need not of twice the force but more than that to give twice the velocity to the same body. No one should be surprised that in ordinary machines, such as the lever, the wheel and axle, the pulley, the wedge, the screw, and the like, there is equilibrium when the size of one body is compensated by the velocity of the other, which is brought about by the arrangement of the machine; or when the magnitudes (the same sort of body being assumed) are reciprocally as the velocities, or when the same quantity of motion is produced in any other way, for then it will happen that there will be the same quantity of effect in both the bodies, or the same height of ascent or descent, on whichever side of the equilibrated system you choose to produce the motion. And so by accident it happens in this case that the force can be estimated from the quantity of motion. But other cases occur, such as that which we have previously dealt with, where they are not the same.

While there is nothing more simple than our demonstration, it is strange that it never came into the minds of Descartes or of his learned followers. But too great confidence in his own ingenuity led him astray, and the others were led astray by their confidence in him. For Descartes, by an error common to great men, became a little too confident. And I fear that not a few of his followers have been imitating the Peripatetics whom they laugh at, that is, they have been depending upon consulting the books of their master, rather than on right reason and the nature of things.

Therefore we may say that forces are in the compound ratio of the bodies (of the same specific gravity or density) and of the heights productive of velocity, that is, the heights by falling from which they can acquire such velocities; or more generally (since no velocity has really been produced) of the heights that will produce them: but not generally of the velocities themselves, although this seems plausible at first sight, and has so seemed to many; from which many errors have arisen, which are contained in the works on mathematical mechanics of RR. PP. Honoratus Faber and Claudius des Chales, and also of Joh. Alph. Borelli and of other men, otherwise distinguished in such matters. Hence also I think it has happened, that doubts have been thrown by some learned men on the theorem of Huygens about the center of oscillation of a pendulum, which however is certainly valid.

On the Interaction of Natural Forces

HERMANN HELMHOLTZ, from *Popular Scientific Lectures*

Hermann Helmholtz (1821–1894) was one of the leading nine-teenth century theoretical physicists and physiologists on the con-tinent. The first to place the conservation of mechanical energy on a sound theoretical basis, Helmholtz showed in his famous paper of 1847 ("The Conservation of Energy") that if the only forces acting between bodies are central forces, it is a consequence of the laws of motion that the total (mechanical) energy is con-served. Thus, the principle of conservation of energy was on as sound a footing as mechanics itself. The demonstration depends upon there being only central forces, that is, forces which act be-tween bodies whose magnitude depends only on the distance be-tween them, and whose direction is along the line which joins them. This result so impressed Helmholtz and others that for a time they believed that the only satisfactory kind of physical expla-nation was one which referred to masses and central forces which acted between them.

But Helmholtz was also aware of other, nonmechanical kinds of causes such as heat, light, electricity, and magnetism, all of which were considered forces. For these other forces Helmholtz offered no conservation theorem. Instead he pointed out in a paper of 1854 ("On the Interaction of Natural Forces") that numerous at-tempts to construct perpetual motion machines using the forces of heat, light, electricity, and magnetism have consistently failed. These failures increased the confidence in the proposition that there are no perpetual motion machines. Once this was accepted as true, its implications had an enormous influence on the turn of scientific problems. It was no longer a problem of using electrical effects on gimmicks to produce magnetic effects which would in turn produce even greater electrical effects. The search for such gimmicks was doomed to failure if there were no perpetual motion machines. The nonexistence of these machines implied, therefore,

Reprinted from *Popular Scientific Lectures*, 1865.

that there were certain relations between quantities of heat, light, electricity and other forces such that it was impossible, for example, by means of a certain quantity of electricity to obtain more than a fixed amount of heat.

Then a new set of problems confronted scientists: what are the specific relations between natural forces? These relations are the so-called interactions of natural forces, and it was by their aid that Helmholtz thought it would be shown that Nature had a total quantity of force which was conserved. This result is no theorem of mechanics. But Helmholtz tried to show why the wider conservation principle is probably true, and he indicated the kind of evidence for which one ought to look.

From these efforts to imitate living creatures, another idea, also by a misunderstanding, seems to have developed itself, one which, as it were, formed the new philosopher's stone of the seventeenth and eighteenth centuries. It was now the endeavour to construct a perpetual motion. Under this term was understood a machine which, without being wound up, without consuming in the working of it falling water, wind, or any other natural force, should still continue in motion, the motive power being perpetually supplied by the machine itself. Beasts and human beings seemed to correspond to the idea of such an apparatus, for they moved themselves energetically and incessantly as long as they lived, and were never wound up; nobody set them in motion. A connexion between the supply of nourishment and the development of force did not make itself apparent. The nourishment seemed only necessary to grease, as it were, the wheelwork of the animal machine, to replace what was used up, and to renew the old. The development of force out of itself seemed to be the essential peculiarity, the real quintessence of organic life. If, therefore, men were to be constructed, a perpetual motion must first be found.

·   ·   ·   ·   ·

. . . Bewildered intellects, however, proclaimed often enough that they had discovered the grand secret; and as the incorrectness of their proceedings was always speedily manifest, the matter fell into bad repute, and the opinion strengthened itself more and more that the problem was not capable of solution; one diffi-

culty after another was brought under the dominion of mathematical mechanics, and finally a point was reached where it could
be proved that at least by the use of pure mechanical forces no
perpetual motion could be generated.

We have here arrived at the idea of the driving force or power
of a machine, and shall have much to do with it in future. I must
therefore give an explanation of it. The idea of work is evidently transferred to machines by comparing their performances
with those of men and animals, to replace which they were applied. . . . Thus the idea of the quantity of work in the case
of machines has been limited to the consideration of the expenditure of force; this was the more important, as indeed most
machines are constructed for the express purpose of exceeding,
by the magnitude of their effects, the powers of men and animals.
Hence, in a mechanical sense, the idea of work has become
identical with that of the expenditure of force, and in this way
I will apply it in the following pages.

How, then, can we measure this expenditure, and compare
it in the case of different machines?

I must here conduct you a portion of the way—as short a
portion as possible—over the uninviting field of mathematico-
mechanical ideas, in order to bring you to a point of view from
which a more rewarding prospect will open. And though the
example which I will here choose, namely, that of a water-mill
with iron hammer, appears to be tolerably romantic, still, alas!
I must leave the dark forest valley, the foaming brook, the
spark-emitting anvil, and the black Cyclops wholly out of sight,
and beg a moment's attention for the less poetic side of the
question, namely, the machinery. This is driven by a water-
wheel, which in its turn is set in motion by the falling water.
The axle of the water-wheel has at certain places small projec-
tions, thumbs, which, during the rotation, lift the heavy hammer
and permit it to fall again. The falling hammer belabours the
mass of metal which is introduced beneath it. The work therefore done by the machine consists, in this case, in the lifting
of the hammer, to do which the gravity of the latter must be overcome. The expenditure of force will, in the first place, other
circumstances being equal, be proportional to the weight of the
hammer; it will, for example, be double when the weight of the
hammer is doubled. But the action of the hammer depends not

upon its weight alone, but also upon the height from which it falls. If it falls through two feet, it will produce a greater effect than if it falls through only one foot. It is, however, clear that if the machine, with a certain expenditure of force, lifts the hammer a foot in height, the same amount of force must be expended to raise a second foot in height. The work is therefore not only doubled when the weight of the hammer is increased twofold, but also when the space through which it falls is doubled. From this it is easy to see that the work must be measured by the product of the weight into the space through which is ascends. And in this way, indeed, we measure in mechanics. The unit of work is a foot-pound, that is, a pound weight raised to the height of one foot.[1]

While the work in this case consists in the raising of the heavy hammer-head, the driving force which sets the latter in motion is generated by falling water. It is not necessary that the water should fall vertically, it can also flow in a moderately inclined bed; but it must always, where it has water-mills to set in motion, move from a higher to a lower position. Experiment and theory concur in teaching that when a hammer of a hundredweight is to be raised one foot, to accomplish this at least a hundredweight of water must fall through the space of one foot; or, what is equivalent to this, two hundredweight must fall half a foot, or four hundredweight a quarter of a foot, &c. In short, if we multiply the weight of the falling water by the height through which it falls, and regard, as before, the product as the measure of the work, then the work performed by the machine in raising the hammer can, in the most favourable case, be only equal to the number of foot-pounds of water which have fallen in the same time. In practice, indeed, this ratio is by no means attained: a great portion of the work of the falling water escapes unused, inasmuch as part of the force is willingly sacrificed for the sake of obtaining greater speed.

I will further remark that this relation remains unchanged whether the hammer is driven immediately by the axle of the wheel, or whether—by the intervention of wheelwork, endless screws, pulleys, ropes—the motion is transferred to the hammer. We may, indeed, by such arrangements succeed in raising a hammer of ten hundredweight, when by the first simple arrangement the elevation of a hammer of one hundredweight might alone be possible; but either this heavier hammer is raised to only one tenth of the height, or tenfold the time is required to raise it to the same height; so that, however we may alter, by the interposition of machinery, the intensity of the acting force, still in a certain time, during which the millstream furnishes us with a definite quantity of water, a certain definite quantity of work, and no more, can be performed.

·   ·   ·   ·   ·

From these examples you observe, and the mathematical theory has corroborated this for all purely mechanical, that is to say, for moving forces, that all our machinery and apparatus generate no force, but simply yield up the power communicated to them by natural forces—falling water, moving wind, or by the muscles of men and animals. After this law had been established by the great mathematicians of the last century, a perpetual motion, which should make use solely of pure mechanical forces, such as gravity, elasticity, pressure of liquids and gases, could only be sought after by bewildered and ill-instructed people. But there are still other natural forces which are not reckoned among the purely moving forces—heat, electricity, magnetism, light, chemical forces, all of which nevertheless stand in manifold relation to mechanical processes. There is hardly a natural process to be found which is not accompanied by mechanical actions, or from which mechanical work may not be derived. Here the question of a perpetual motion remained open; the decision of this question marks the progress of modern physics, regarding which I promised to address you.

·   ·   ·   ·   ·

But, warned by the futility of former experiments, the public had become wiser. On the whole, people did not seek much after combinations which promised to furnish a perpetual motion, but the question was inverted. It was no more asked, How can I make use of the known and unknown relations of natural forces so as to construct a perpetual motion? but it was asked, If a perpetual motion be impossible, what are the relations which must subsist between natural forces? Everything was gained by this inversion of the question. The relations of natural forces, rendered necessary by the above assumption, might be easily and completely stated. It was found that all known relations of forces harmonise with the consequences of that assumption, and a series of unknown relations were discovered at the same time, the correctness of which remained to be proved. If a single one of them could be proved false, then a perpetual motion would be possible.

The first who endeavoured to travel this way was a Frenchman named Carnot, in the year 1824. In spite of a too limited conception of his subject, and an incorrect view as to the nature of heat which led him to some erroneous conclusions, his experiment was not quite unsuccessful. He discovered a law which now bears his name, and to which I will return further on.

His labours remained for a long time without notice, and it was not till eighteen years afterwards, that is in 1842, that different investigators in different countries, and independent of Carnot, laid hold of the same thought. The first who saw truly the general law here referred to, and expressed it correctly, was a German physician, J. R. Mayer of Heilbronn, in the year 1842. A little later, in 1843, a Dane named Colding presented a memoir to the Academy of Copenhagen, in which the same law found utterance, and some experiments were described for its further corroboration. In England, Joule began about the same time to make experiments having reference to the same subject. We often find, in the case of questions to the solution of which the development of science points, that several heads, quite independent of each other, generate exactly the same series of reflections.

I myself, without being acquainted with either Mayer or Colding, and having first made the acquaintance of Joule's experiments at the end of my investigation, followed the same path. I endeavoured to ascertain all the relations between the different natural processes, which followed from our regarding them from the above point of view. My inquiry was made public in 1847, in a small pamphlet bearing the title, "On the Conservation of Force."

Since that time the interest of the scientific public for this subject has gradually augmented, particularly in England, of which I had an opportunity of convincing myself during a visit last summer. A great number of the essential consequences of the above manner of viewing the subject, the proof of which was wanting when the first theoretic notions were published, have since been confirmed by experiment, particularly by those of Joule; and during the last year the most eminent physicist of France, Regnault, has adopted the new mode of regarding the question, and by fresh investigations on the specific heat of gases has contributed much to its support. For some important consequences the experimental proof is still wanting, but the number of confirmations is so predominant, that I have not deemed it premature to bring the subject before even a non-scientific audience.

How the question has been decided you may already infer from what has been stated. In the series of natural processes there is no circuit to be found, by which mechanical force can be gained without a corresponding consumption. The perpetual motion remains impossible. Our reflections, however, gain thereby a higher interest.

We have thus far regarded the development of force by natural processes, only in its relation to its usefulness to man, as mechanical force. You now see that we have arrived at a general law, which holds good wholly independent of the application which man makes of natural forces; we must therefore make the expression of our law correspond to this more general significance. It is in the first place clear, that the work which, by any natural process whatever, is performed under favourable conditions by a machine, and which may be measured in the way already indicated, may be used as a measure of force common to all. Further, the important question arises, If the quantity of force cannot be augmented except by corresponding consumption, can it be diminished or lost? For the purposes of our machines it certainly can, if we neglect the opportunity to convert natural processes to use, but as investigation has proved, not for nature as a whole.

In the collision and friction of bodies against each other, the mechanics of former years assumed simply that living force was lost. But I have already stated that each collision and each act of friction generates heat; and, moreover, Joule has established by experiment the important law, that for every footpound of force which is lost a definite quantity of heat is always generated, and that when work is performed by the consumption of heat, for each foot-pound thus gained a definite quantity of heat disappears. The quantity of heat necessary to raise the temperature of a pound of water a degree of the Centigrade thermometer, corresponds to a mechanical force by which a pound weight would be raised to the height of 1,350 feet: we name this quantity the mechanical equivalent of heat. I may mention here that these facts conduct of necessity to the conclusion, that heat is not, as was formerly imagined, a fine imponderable substance, but that, like light, it is a peculiar shivering motion of the ultimate particles of bodies. In collision and friction, according to this manner of viewing the subject, the motion of the mass of a body which is apparently lost is converted into a motion of the ultimate particles of the body; and conversely, when mechanical force is generated by heat, the motion of the ultimate particles is converted into a motion of the mass.

.   .   .   .

From a similar investigation of all the other known physical and chemical processes, we arrive at the conclusion that Nature as a whole possesses a store of force which cannot in any way be either increased or diminished, and that therefore the quantity of force in Nature is just as eternal and unalterable as the quantity of matter. Expressed in this form, I have named the general law "The Principle of the Conservation of Force."

Conservation and Causality

MICHAEL FARADAY, *The Conservation of Force*

Michael Faraday (1791–1867), superb experimental physicist that he was, seems to have caught the feelings of both experimental and theoretical physicists alike when he described his attitude to the conservation of force. Faraday meant by "force" any cause of a physical action. It is apparent, therefore, that the conservation of force, according to his understanding of it, was not restricted to the domain of mechanics. He thought that the conservation of force requires the constancy of a physical quantity over time and despite various interactions. This conserved quantity, moreover, was not any old quantity; it had to be an important one, and Faraday thought of it as playing a causal role in explanations of physical change.

At the time Faraday wrote on the subject of conservation, Helmholtz had already shown (1847) that under certain conditions the mechanical energy of some systems of bodies is conserved. Also, Joule's work on the mechanical equivalent of heat was known (see Joule, p. 204), so that Faraday thought there was good evidence for the conservation of force. But Faraday and others did more than believe in the truth of the principle; they regarded the conservation of force as so important that they made it central to the methodology of science. Faraday tells us that no hypothesis ought to be considered if it is incompatible with the conservation of force. Only those statements are to be investigated, tested, or otherwise taken seriously by scientists which are compatible with the conservation principle. In addition, the principle suggests whether further work is required in a science. For example, suppose that a quantity Q was appealed to in certain causal explanations, but it was not conserved. Faraday thought that one ought to look for further quantities $Q_1, Q_2, \ldots, Q_n$ and laws relating the Q's to each other, such that all quantities contribute to, or

Reprinted from *Philosophical Magazine*, IV, April, 1857.

constitute, a force which is conserved, while all the individual quantities are just so many different forms of that one force.

Finally, Faraday was well aware of the empirical basis of the conservation of force. Even though he argued for its centrality and perhaps dominance in physical inquiry, he also cautioned that the principle might one day be shown false. He did not rule out this possibility, but remarked only that if the law proved false, we could look forward to some very amazing discoveries.

There is no question which lies closer to the root of all physical knowledge than that which inquires whether force can be destroyed or not. The progress of the strict science of modern times has tended more and more to produce the conviction that "force can neither be created nor destroyed"; and to render daily more manifest the value of the knowledge of that truth in experimental research. To admit, indeed, that force may be destructible or can altogether disappear, would be to admit that matter could be uncreated; for we know matter only by its forces; and though one of these is most commonly referred to, namely, gravity, to prove its presence, it is not because gravity has any pretension, or any exemption, amongst the forms of force as regards the principle of *conservation,* but simply that being, as far as we perceive, inconvertible in its nature and unchangeable in its manifestation, it offers an unchanging test of the matter which we recognize by it.

Agreeing with those who admit the conservation of force to be a principle in physics, as large and sure as that of the indestructibility of matter, or the invariability of gravity, I think that no particular idea of force has a right to unlimited or unqualified acceptance that does not include *assent* to it; and also, to *definite amount* and *definite disposition of the force,* either in one effect or another, for these are necessary consequences; therefore I urge, that the conservation of force ought to be admitted as a physical principle in all our hypotheses, whether partial or general, regarding the actions of matter. I have had doubts in my own mind whether the considerations I am about to advance are not rather metaphysical than physical. I am unable to define what is metaphysical in physical science; and am exceedingly adverse to the easy and unconsidered admission of one supposition upon another, suggested as they often are by very imperfect induction from a small number of facts, or by a very imperfect observation of the facts themselves; but, on the other hand, I think the philosopher may be bold in his application of principles which have been developed by close inquiry, have stood through much investigation, and continually increase in force. For instance, *time* is growing up daily into importance as an element in the exercise of force. The earth moves in its orbit in time; the crust of the earth moves in time; light moves in time; an electromagnet requires time for its charge by an electric current; to inquire, therefore, whether power, acting either at sensible distances, always acts in *time,* is not to be metaphysical; if it acts in time and across space, it must act by physical lines of force; and our view of the nature of the force may be affected to the extremest degree by the conclusions which experiment and observation on time may supply; being, perhaps, finally determinable only by them. To inquire after the possible time in which gravitating, magnetic, or electric force is exerted, is no more metaphysical than to mark the times of the hands of a clock in their progress; or that of the temple of Serapis and its ascents and descents; or the periods of the occultations of Jupiter's satellites; or that in which the light from them comes to the earth. Again, in some of the known cases of action in time, something happens whilst the *time* is passing which did not happen before, and does not continue after; it is, therefore, not metaphysical to expect an effect in *every* case, or to endeavour to discover its existence and determine its nature. So in regard to the principle of the conservation of force; I do not think that to admit it, and its consequences, whatever they may be, is to be metaphysical; on the contrary, if that word have any application to physics, then I think that any hypothesis, whether of heat, or electricity, or gravitation, or any other form of force, which either willingly or unwillingly dispenses with the principle of conservation, is more liable to the charge than those which, by including it, become so far more strict and precise.

Supposing that the truth of the principle of the conservation of force is assented to, I come to *its uses.* No hypothesis should be admitted, nor any assertion of a fact credited, that denies the principle. No view should be inconsistent or incompatible with it. Many of our hypotheses in the present state of science may not comprehend it, and may be unable to suggest its consequences; but none should oppose or contradict it.

If the principle be admitted, we perceive at once that a theory or definition, though it may not contradict the principle, cannot be accepted as sufficient or complete unless the former be contained in it; that however well or perfectly the definition may include and represent the state of things commonly considered under it, that state or result is only partial, and must not be accepted as exhausting the power or being the full equivalent, and therefore cannot be considered as representing its *whole nature;* that indeed, it may express only a very small part of the whole, only a residual phenomenon, and hence give us but little indication of the full natural truth. Allowing the principle its force, we ought, in every hypothesis, either to account for its consequences by saying what the changes are when force of a given kind apparently disappears, as when ice thaws, or else should leave space for the idea of the conversion. If any hypothesis, more or less trustworthy on other accounts, is insufficient in expressing it or incompatible with it, the place of deficiency or opposition should be marked as the most important for examination, for there lies the hope of a discovery of new laws or a new condition of force. The deficiency should never be accepted as satisfactory, but be remembered and used as a stimulant to further inquiry; for conversions of force may here be hoped for. Suppositions may be accepted for the time, provided they are not in contradiction with the principle. Even an increased or diminished capacity is better than nothing at all, because such a supposition, if made, must be consistent with the nature of the original hypothesis, and may, therefore, by the application of experiment, be converted into a further test of probable truth. The case of a force simply removed or suspended, without a transferred exertion in some other direction, appears to me to be absolutely impossible.

If the principle be accepted as true, we have a right to pursue it to its consequences, no matter what they may be. It is, indeed, a duty to do so. A theory may be perfection, as far as it goes, but a consideration going beyond it, is not for that reason to be shut out. We might as well accept our limited horizon as the limits of the world. No magnitude, either of the phenomena or of the results to be dealt with, should stop our exertions to ascertain, by the use of the principle, that something remains to be discovered, and to trace in what direction that discovery may lie.

I will endeavour to illustrate some of the points which have been urged, by reference, in the first instance, to a case of power, which has long had great attractions for me, because of its extreme simplicity, its promising nature, its universal presence, and in its invariability under like circumstances; on which, though I have experimented and as yet failed, I think experiment would be well bestowed, I mean the force of gravitation. I believe I represent the received idea of the gravitating force aright in saying that it is *a simple attractive force exerted between any two or all the particles or masses of matter, at every sensible distance, but with a strength varying inversely as the square of the distance*. The usual idea of the force implies *direct* action at a distance; and such a view appears to present little difficulty except to Newton, and a few, including myself, who in that respect may be of like mind with him.

This idea of gravity appears to me to ignore entirely the principle of the conservation of force; and by the terms of its definition, if taken in an absolute sense, *"varying* inversely as the square of the distance," to be in direct opposition to it, and it becomes my duty to point out where this contradiction occurs, and to use it in illustration of the principle of conservation. Assume two particles of matter, A and B, in free space, and a force in each or in both by which they gravitate towards each other, the force being unalterable for an unchanging distance, but varying inversely as the square of the distance when the latter varies. Then, at the distance of ten, the force may be estimated as one; whilst at the distance of one, that is, one-tenth of the former, the force will be one hundred; and if we suppose an elastic spring to be introduced between the two as a measure of the attractive force, the power compressing it will be a hundred times as much in the latter case as in the former. But from whence can this enormous increase of power come? If we say that it is the character of this force, and content ourselves with that as a sufficient answer, then it appears to me we admit a *creation* of power and that to an enormous amount; yet by a change of condition, so small and simple as to fail in lending the least instructed mind to think that it can be sufficient cause, we should admit a result which would equal the highest act our minds can appreciate of the working of infinite power upon matter; we should let loose the highest law in physical science which our faculties permit us to perceive, namely, the *conservation of force*.

Suppose the two particles, A and B, removed back to the greater distance of ten, then the force of attraction would be only a hundredth part of that they previously possessed; this, according to the statement that the force varies inversely as the square of the distance, would double the strangeness of the above results; it would be an *annihilation of force*—an effect equal in its infinity and its consequences with *creation,* and only within the power of Him who has created.

We have a right to view gravitation under every form that either its definition or its effects can suggest to the mind; it is our privilege to do so with every force in nature; and it is only by so doing that we have succeeded, to a large extent, in relating the various forms of power, so as to derive one from another, and thereby obtain confirmatory evidence of the great principle of the conservation of force.

The principle of the conservation of force would lead us to assume, that when A and B attract each other less, because of increasing distance, then some other exertion of power, either within or without them, is proportionately growing up; and again, that when their distance is diminished, as from ten to one, the power of attraction, now increased a hundred-fold, has been produced out of some other form of power which has been equivalently reduced. This enlarged assumption of the nature of gravity is not more metaphysical than the half assumption; and is, I believe, more philosophical and more in accordance with all physical considerations. The half assumption is, in my view of the matter, more dogmatic and irrational than the whole, because it leaves it to be understood that power can be created and destroyed almost at pleasure.

The principles of physical knowledge are now so far developed as to enable us not merely to define or describe the *known,* but to state reasonable expectations regarding the *unknown;* and I think the principle of the conservation of force may greatly aid experimental philosophers in that duty to science, which consists in the enunciation of problems to be solved. It will lead us, in any case where the force remaining unchanged in form is altered in direction only, to look for the new disposition of the force; as in the cases of magnetism, static electricity, and perhaps gravity, and to ascertain that as a whole it remains unchanged in amount—or, if the original force disappear, either altogether or in part, it will lead us to look for the new condition or form of force which should result, and to develop its equivalency to the force that has disappeared. Likewise, when force is developed, it will cause us to consider the previously-existing equivalent to the force so appearing; and many such cases there are in chemical action. When force disappears, as in the electric or magnetic induction after more or less discharge, or that of gravity with an increasing distance, it will suggest a research as to whether the equivalent change is one within the apparently acting bodies, or one *external* (in part) to them. It will also raise up inquiry as to the nature of the internal or external state, both before the change and after. If supposed to be external, it will suggest the necessity of a physical process, by which the power is communicated from body to body; and in the case of external action, will lead to the inquiry whether, in any case, there can be truly action at a distance, or some other medium, is not necessarily present.

But after all, the principle of the conservation of force may by some be denied. Well, then, if it be unfounded even in its application to the smallest part of the science of force, the proof must be within our reach, for all physical science is so. In that case, discoveries as large or larger than any yet made, may be anticipated. I do not resist the search for them, for no one can do harm, but only good, who works with an earnest and truthful spirit in such a direction. But let us not admit the destruction or creation of force without clear and constant proof. Just as the chemist owes all the perfection of his science to his dependence on the certainty of gravitation applied by the balance, so may the physical philosopher expect to find the greatest security and the utmost aid in the principle of the conservation of force. All that we have that is good and safe, as the steam-engine, the electric-telegraph, &c., witness to that principle—it would require a perpetual motion, a fire without heat, heat without a source, action without reaction, cause with effect, or effect without a cause, to displace it from its rank as a law of nature.

The Location of Energy

J. CLERK MAXWELL, *Letter to P. G. Tait*

In his comments upon the conservation of charge (Maxwell, p. 217), Maxwell 'expressed doubt whether energy was conserved if there were action at a distance. If a charged body could influence another some distance from it without benefit of any intervening medium, then energy was probably not conserved.

Maxwell had several reasons for believing that such a medium existed. First, his approach to electrodynamics, shared with and inspired by Faraday, stressed that the interaction of charged bodies gave insight into the nature of an intervening medium. Second, as is evident in this letter to P. G. Tait (a physicist famous in his own right), he thought there were only two options open to anyone who wished to construct a theory of electrodynamic action. Either a certain medium existed in space such that disturbances moved through it according to certain laws, or there was action at a distance. The first alternative he took to be the core of his theory of electrodynamics. The specific laws of propagation which he proposed are now referred to as Maxwell's equations. If two bodies were apart, their interaction was to be explained by showing how the behavior of one of them caused a certain disturbance in the medium. The disturbance then propagated according to Maxwell's equations to the vicinity of the second body, where it then produced its effect. He believed that the medium or field was an ineliminable part of the explanation because the second alternative, action at a distance, led, he thought, to a case in which a rigid rod seems to "pull itself along" perpetually—an apparent violation of the principle of energy conservation. He therefore maintained the correctness of his own approach to electrodynamic theory. Using the laws of propagation of the medium—the Maxwell equations—he tried to show that the energy of a system could be located not only in its bodies, but also in the field which contained them.

Maxwell, *Letter to P. G. Tait,* 12 March 1868. Unpublished letter in archives of the Rayleigh Library, Cavendish Laboratories, Cambridge, England. Printed by permission of the Library, and Brigadier John Wedderburn-Maxwell.

8 P.G.T.

Dr Tait   12 March 1868

. . . . .

With respect to Riemann for whom I have great respect and regret I only lately got ether Pogs or Phil Mag from the binder & wrote you a rough note for yourself. I now have him more distinct. Weber says that electrical force depends on the distance and its 1st & 2nd derivatives with respect to t.

Riemann says that this is due to the fact that the potential at a point is due to the distribution of electricity elsewhere not at that instant but at times before depending on the distances.

In other words potential is propagated through space at a certain rate and he actually expresses this by a partial diff eqn appropriate to propagation. Hence either (1) space contains a medium capable of dynamical actions which go on during transmission independently of the causes which excited them (and this is no more or less than my theory divested of particular assumptions) or (2) if we consider the hypothesis as a fact without any etherial substratum

$$\overset{\displaystyle X}{\underset{\rightarrow}{\cdot}} \qquad \overset{\displaystyle Y}{\underset{\rightarrow}{\cdot}}$$

Let X & Y be travelling to the right with velocity v at a distance a then the force of X on Y will be $\dfrac{XY}{a^2}\left(1-\dfrac{v}{V}\right)^2$ and that of Y on X $\dfrac{XY}{a^2}(1+v/V)^2$ Where V is the velocity of transmission of force. If the force is an attraction and if X & Y are connected by a rigid rod X will be pulled forward more than Y is pulled back and the system will be a locomotive engine fit to carry you through space with continually increasing velocity. See Gullivers Travels in Laputa.

. . . . .

Riemann's action & reaction between the gross bodies are unequal and his energy is nowhere unless he admits a medium which he does not do explicitly. My action & reaction are equal only between things in contact not between the gross bodies till they have been in position for a sensible time, and my energy is and remains in the medium including the gross bodies which are among it.

Yours truly

J Clerk Maxwell

Conservation of Mass

ALBERT EINSTEIN, $E = MC^2$

The equation $E = mc^2$ is one of the most widely known among professional scientists and laymen alike. The equation which relates the energy of a physical system and its mass is, as Einstein pointed out in his article, inexact, but startling enough even in its approximate form. According to the Special Theory of Relativity, the exact relation states that the total energy of a system is equal to the product of three factors: the rest mass m (that is, the mass as measured by an observer moving with the system); c^2, the square of the velocity of light (*in vacuo*); and a third quantity, $1/\sqrt{1 - (v^2/c^2)}$, which is a function of the velocity of the system. The important feature of the last quantity is that it is equal to 1 when the velocity is zero. The remarkable consequence of this exact relation is that even if a body is at rest, then it still has a (rest) energy E equal to mc^2. According to classical theories of motion, bodies have energy either because of their position (potential energy) or their motion (kinetic energy). But for the first time we learn that there is a definite energy that a body has which is neither an energy of position nor motion, but seems to be an incredibly large energy associated with the body itself.

The law relating energy and mass also serves to blend two principles of conservation (see "Introduction," Part II). We have already noted that after it was discovered that heat was not conserved, Joule found that heat and mechanical work could be converted into each other. The Special Theory of Relativity taught that in certain processes the mass is not conserved, contrary to what had been believed ever since modern science began. Nevertheless Einstein showed that there was a lawlike connection between energy and mass, and each can be presumably converted to the other. In the case of heat, like mass, a principle of conservation was found incorrect. But a theory supplied a new relation between those quantities and others. With the aid of the new law, one could explain why the conservation of mass was sound only under

Reprinted by permission from *Science Illustrated,* April, 1946.

247

some conditions and not others. This theoretical advance has been described as the blending of principles of conservation. Heat and mechanical work were blended by the results of Joule, and mass became a form of energy as a result of Einstein's work.

In order to understand the law of the equivalence of mass and energy, we must go back to two conservation or "balance" principles which, independent of each other, held a high place in pre-relativity physics. These were the principle of the conservation of energy and the principle of the conservation of mass. The first of these, advanced by Leibnitz as long ago as the seventeenth century, was developed in the nineteenth century essentially as a corollary of a principle of mechanics.

Consider, for example, a pendulum whose mass swings back and forth between the points A and B. At these points the mass m is higher by the amount h than it is at C, the lowest point of the path (see drawing [Fig. 1]). At C, on the other hand, the

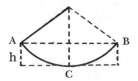

Fig. 1   DRAWING FROM DR. EINSTEIN'S MANUSCRIPT

lifting height has disappeared and instead of it the mass has a velocity v. It is as though the lifting height could be converted entirely into velocity, and vice versa. The exact relation would be expressed as $mgh = \frac{m}{2}v^2$, with g representing the acceleration of gravity. What is interesting here is that this relation is independent of both the length of the pendulum and the form of the path through which the mass moves.

The significance is that something remains constant throughout the process, and that something is energy. At A and at B it is an energy of position, or "potential" energy; at C it is an energy of motion, or "kinetic" energy. If this concept is correct, then the sum $mgh + m\frac{v^2}{2}$ must have the same value for any position of the pendulum, if h is understood to represent the height above

C, and v the velocity at that point in the pendulum's path. And such is found to be actually the case. The generalization of this principle gives us the law of the conservation of mechanical energy. But what happens when friction stops the pendulum?

The answer to that was found in the study of heat phenomena. This study, based on the assumption that heat is an indestructible substance which flows from a warmer to a colder object, seemed to give us a principle of the "conservation of heat." On the other hand, from time immemorial it has been known that heat could be produced by friction, as in the fire-making drills of the Indians. The physicists were for long unable to account for this kind of heat "production." Their difficulties were overcome only when it was successfully established that, for any given amount of heat produced by friction, an exactly proportional amount of energy had to be expended.[1] Thus did we arrive at a principle of the "equivalence of work and heat." With our pendulum, for example, mechanical energy is gradually converted by friction into heat.

In such fashion the principles of the conservation of mechanical and thermal energies were merged into one. The physicists were thereupon persuaded that the conservation principle could be further extended to take in chemical and electromagnetic processes—in short, could be applied to all fields. It appeared that in our physical system there was a sum total of energies that remained constant through all changes that might occur.

Now for the principle of the conservation of mass. Mass is defined by the resistance that a body opposes to its acceleration (inert mass). It is also measured by the weight of the body (heavy mass). That these two radically different definitions lead to the same value for the mass of a body is, in itself, an astonishing fact. According to the principle—namely, that masses remain unchanged under any physical or chemical changes—the mass appeared to be the essential (because unvarying) quality of matter. Heating, melting, vaporization, or combining into chemical compounds would not change the total mass.

Physicists accepted this principle up to a few decades ago. But it proved inadequate in the face of the special theory of relativity. It was therefore merged with the energy principle—just as, about sixty years before, the principle of the conservation of mechanical energy had been combined with the principle of the conservation of heat. We might say that the principle of the conservation of energy, having previously swallowed up that of the conservation of heat, now proceeded to swallow that of the conservation of mass—and holds the field alone.

It is customary to express the equivalence of mass and energy (though somewhat inexactly) by the formula $E = mc^2$, in which c represents the velocity of light, about 186,000 miles per second. E is the energy that is contained in a stationary body; m is its mass. The energy that belongs to the mass m is equal to this mass, multiplied by the square of the enormous speed of light— which is to say, a vast amount of energy for every unit of mass.

But if every gram of material contains this tremendous energy, why did it go so long unnoticed? The answer is simple enough: so long as none of the energy is given off externally, it cannot be observed. It is as though a man who is fabulously rich should never spend or give away a cent; no one could tell how rich he was.

Now we can reverse the relation and say that an increase of E in the amount of energy must be accompanied by an increase of $\dfrac{E}{c^2}$ in the mass. I can easily supply energy to the mass—for instance, if I heat it by ten degrees. So why not measure the mass increase, or weight increase, connected with this change? The trouble here is that in the mass increase the enormous factor c^2 occurs in the denominator of the fraction. In such a case the increase is too small to be measured directly; even with the most sensitive balance.

For a mass increase to be measurable, the change of energy per mass unit must be enormously large. We know of only one sphere in which such amounts of energy per mass unit are released: namely, radioactive disintegration. Schematically, the process goes like this: An atom of the mass M splits into two atoms of the mass M' and M'', which separate with tremendous kinetic energy. If we imagine these two masses as brought to rest —that is, if we take this energy of motion from them—then, considered together, they are essentially poorer in energy than was the original atom. According to the equivalence principle, the mass sum $M' + M''$ of the disintegration products must also be somewhat smaller than the original mass M of the disintegrating atom—in contradiction to the old principle of the conservation of mass. The relative difference of the two is on the order of one-tenth of one percent.

Now, we cannot actually weigh the atoms individually. However, there are indirect methods for measuring their weights exactly. We can likewise determine the kinetic energies that are transferred to the disintegration products M' and M''. Thus it has become possible to test and confirm the equivalence formula. Also, the law permits us to calculate in advance, from precisely determined atomic weights, just how much energy will be released with any atomic disintegration we have in mind. The law says nothing, of course, as to whether—or how—the disintegration reaction can be brought about.

What takes place can be illustrated with the help of our rich man. The atom M is a rich miser who, during his life, gives away no money (*energy*). But in his will he bequeaths his fortune to his sons M' and M'', on condition that they give to the community a small amount, less than one-thousandth of the whole estate (*energy or mass*). The sons together have somewhat less than the father had (*the mass sum $M'+M''$ is somewhat smaller than the mass M of the radioactive atom*). But the part given to the community, though relatively small, is still so enormously large (*considered as kinetic energy*) that it brings with it a great threat of evil. Averting that threat has become the most urgent problem of our time.

The Principles of Conservation

MAX VON LAUE, *Inertia and Energy*

This very beautiful essay on the principles of conservation has received just praise from all quarters. Von Laue (1879–1960), himself a Nobel Laureate, contributed to the development of the Special Theory of Relativity by his early investigation of the experimental grounds for the theory as well as his theoretical investigation of the implications of Einstein's theory for the energy and momentum radiated by a special system of moving charged bodies.

Von Laue begins with a description of some of the most important principles of conservation in classical physics. He arranges them into a hierarchy and gives a brief history of the discovery and development of each.

But two principles are more fundamental than the rest: the Law of (linear) Inertia, which is simply the conservation of linear momentum (the product of the mass and velocity of a body), and the conservation of energy. These two principles are related or fused together if we assume a classical principle of relativity which states that the laws of mechanics cannot be used to distinguish between one coordinate system and another which is moving uniformly with respect to it. This principle seems obviously incorrect because we can distinguish between one coordinate system and another moving uniformly with respect to us. But the principle of *classical* relativity is not a denial of this fact; it states that whether we refer our laws of motion to the one coordinate system or the other will make no difference to the classical Newtonian laws of motion—they will have exactly the same form no matter which reference frame is employed. The relating of classical inertia and energy by the principle of classical relativity has great significance for von Laue.

He explains how the pressure of new experimental results and

Reprinted by permission of the publishers from *Albert Einstein: Philosopher-Scientist,* edited by Paul Schilpp, The Library of Living Philosophers, now published by The Open Court Publishing Company, La Salle, Illinois. First edition, 1949.

the emergence of new successful theories strongly suggested the extension of the concepts of inertia and energy beyond their original sphere of mechanics. The new concepts of inertia and energy are not identical with their mechanical counterparts. Yet their use in the description of old as well as new experimental findings brings out their resemblance to the older notions. There is a theoretical similarity, too, by which von Laue is much impressed: in classical mechanics, the principle of conservation of energy is related to the Law of Inertia, given the classical principle of Newtonian relativity. So, too, according to Einstein, the (extended) Law of Inertia and the conservation of energy are deductively related, given Einstein's Principle of Relativity (see Einstein, p. 155).

However, one should not overemphasize the continuity or similarity of the concepts involved, for there are extremely noteworthy differences. The example of the floating cylinder discussed by von Laue shows that, according to the Special Theory of Relativity, if a certain amount of energy ΔE (as heat, γ radiation, or in any other form) is added to a physical system, there is a change in the mass of the system by an amount $(\Delta E)/c^2$. This is truly a striking implication because it states that there can be an increase in the mass of a physical system without any new matter or bodies being added. This is a great departure from the classical Newtonian concept of mass, which served in part to indicate the presence of bodies.

I. Introduction

In modern physics the laws of conservation are of fundamental importance. Essentially, there are three of them: The principle of inertia, which states the conservation of linear momentum; the energy principle, which asserts the conservation of energy; and the law of the conservation of the quantity of electricity. There are two more conservation laws, dealing with the conservation of the angular momentum and of the inert mass. The first of these two, however, is a necessary consequence of the law of conservation of linear momentum, and the latter, as far as we still acknowledge it to be correct, has become identical with the law of the conservation of energy. Finally the laws of conservation of linear momentum and of energy fuse into one for modern relativistic considerations. The discussion that follows is concerned with these unifications of originally distinct laws.

Let us first consider the conservation of the quantity of electricity. The history of this law is soon told, though it extends over a century or more. Under the tremendous impact of Newton's law of gravitational attraction, this law was hypothetically postulated for the electrical attraction and repulsion in the early 18th century. The conservation of mass in which one believed with good reason, was transferred to the electric "Fluida," with the only modification that positive and negative charges cancel each other. Several investigators deduced the inverse proportionality to the square of the distance a long time before Coulomb from the screening effect of electrically conducting enclosures. Coulomb himself cites this effect in a paper on his famous experiment with the torsion balance as a second proof of this law. Only a few persons, however, understood this argument at the time; thus it was forgotten. The proportionality between forces and charges was accepted implicitly by all scientists of that time. They could not have proved it, if for no other reason than that they did not possess a well-defined quantitative measure of the charge. The idea of defining the quantity of electricity from Coulomb's law itself was originated by Carl Friedrich Gauss (about 1840). And the experimental proof of charge conservation was first given by the ice-bucket experiment of Michael Faraday in 1843.

The apparatus consisted of a metal vessel with a relatively small opening, the ice-bucket. This vessel was connected to an electrometer, but was othewise insulated. Faraday lowered an electrically charged body into it, suspended by an insulating string; immediately the electrometer showed a deflection proportional to the charge. And this deflection remained unchanged, no matter what was done with the charge inside the ice-bucket, e.g., whether it was transferred to the wall, or whether it was added to other charges present before the start of the experiment, which might perhaps compensate for the new charge. Perhaps this proof is not too accurate. As an integrating component of Maxwell's theory of electricity, however, the law of conservation of charge is supported by the numerous and exact experimental confirmations of this theory. Nowadays nobody doubts it.

This fundamental result has never caused even approximately the same sensation as the law of the conservation of energy, which was proved at nearly the same time. History makes us understand this difference. Generations of scientists have striven for it, but the conservation of the quantity of charge corresponded to the *"communis opinio"* already a hundred years before Faraday.

In what follows we shall relate the much more eventful story of the two other conservation laws and their interrelationship.

II. The Law of Momentum in Newton's Mechanics

The laws of conservation of momentum and energy are results of modern times. The law of momentum is probably the earlier of the two, because it was clearly formulated, and its significance recognized, sooner than the law of energy, even when we think only of the law of energy in mechanics. It is true, antiquity has left us permanently valuable knowledge for statics, the science of mechanical equilibrium. But influenced by the doctrines of Aristotle, it had advocated a proposition diametrically opposed to the principle of inertia, namely, that a permanent action from without is necessary for maintaining any motion. It was not until the time of Galileo Galilei (1564–1642), that this brilliant originator of the theory of motion, dynamics, in his long life of research, realized that the motion of a body free of outside interaction does not stop, but continues for all time with constant velocity. It is significant that the consideration which led him to this result is based on energy considerations in the modern sense.

It runs as follows: a body on the surface of the earth which falls from a certain altitude, be it directly, on an inclined plane, on the circular path of a pendulum, or otherwise, must obtain precisely that velocity which it requires to return to its former level. For any deviation from this law would furnish a method for making the body ascend by means of its own gravity, perhaps through the inversion of the process of motion which is always possible; and Galileo thinks that this is impossible. Moreover, he confirms his conviction by means of certain experiments. Now let a body ascend again on an inclined plane after it has fallen downwards a certain distance. The lower the inclination toward the horizontal, the longer the path which it requires to obtain its former level on the inclined plane. And if the plane is horizontal, then the body will keep on flying (on it) to infinity with undiminished velocity.

This transition to the limit is not stated explicitly in Galileo's papers, because this "rigorous empiricist" knew that a plane is not a level surface, because of the spherical shape of the earth, and he avoided intentionally any hypothesis concerning conditions as they might be encountered somewhere away from the earth. But the reader of his *Discorsi* on the mechanics (1638) could not help drawing this conclusion himself. Thus the simplest form of the law of inertia, that the velocity of each force-free body is maintained with respect to direction and magnitude, has become the common property of all scientists since Galileo.

It was soon observed, however, that a more general law was hidden in the law of inertia, when the interaction of two bodies was considered. Collisions were then considered to be the simplest kind of interaction; thus even before Newton a whole series of collision theories had arisen. Few of them were based on exact experiments, but all of them were in agreement that the masses of the colliding bodies were important, their masses simply being identified with the weights. As weighing never showed changes in the weight of a body, the mass, too, was considered to be constant. This important step was taken apparently without misgiving, as a matter of course. Incidentally, these earliest collision theories considered only the central impact of two balls, where all motions take place on a straight line (the connecting line of the mid-points).

From the mass m and the velocity q René Descartes (1596–1650) already formed the *quantity of motion mq* and asserted the conservation of the sum of the quantities of motion of all bodies, and in particular of two colliding bodies, on the basis of philosophic-theological speculations which to us appear strange. But he understood the velocity merely as a number, as a scalar quantity in our terminology, without considering its direction, i.e., its vector properties. Thus his considerations naturally did not lead to successful results.

In 1668 the Royal Society in London made the theory of collisions the topic for a contest. Three papers were submitted. The first candidate, John Wallis (1616–1703), well-known in mathematics by the "Wallis formula," observed, indeed, that opposite velocities had to be provided with opposite signs when the above-mentioned quantity of motion is formed. Otherwise he retained the Cartesian idea of the constancy of the quantity of motion; but of course he could not determine the two momenta after an impact (considered by him as in-elastic) from the initial velocities with the help of this single requirement only; he, therefore, introduces additional assumptions which, being incorrect, vitiate his result.

Christian Huygens (1629–1695) first submitted to the Royal Society his results only, but he gave the proofs later in a paper (*De motu corporum ex percussione*) published in his *Opuscula posthuma* in 1703. He realized correctly that not only the sum (formed with the correct signs) of the quantities of motion $m \cdot q$ has the same value before and after a perfectly elastic collision, but also the sum of the products of the respective masses m by the squares of the associated velocities, $m \cdot q^2$. These two statements are in fact sufficient to solve the problem. Here we encounter the first application of the principle of mechanical energy, though without any realization of its comprehensive significance. By the way, Christopher Wren (1632–1723), the third candidate, used the principle in the same manner. It is very interesting, however, that Huygens used in his paper the principle of relativity, of course only the one which corresponds to the mechanics of his time and which we call today the Galilean principle of relativity to distinguish it from that of Einstein. In order to generalize the law that two equal elastic balls with equal velocities and opposite signs simply exchange their velocities on impact, Huygens assumes that the impact occurs on a moving ship, and he observes it from shore. Thus he transforms, as we should call it today, from one system of reference to another.

But all these investigations became dated when Isaac Newton (1642–1727) published his *"Principia"* (*Philosophiæ Naturalis Principia Mathematica*) in 1687. In this treatise we find the two pronouncements that the rate of change of (linear) momentum of a mass point per unit time equals the force acting on it, and that the forces between two mass points are equal and opposite (equality of action and reaction). It follows immediately that the interaction of an arbitrary number of mass points never changes their total momentum, but that this total momentum is constant for any system not subject to external forces. Newton calculates the total momentum as a vector quantity, by adding vectorially the individual momenta. His formulation of the first law above appears almost prophetic: He equates the force not to the product of mass and acceleration, but to the rate of change of momentum, even though both formulations are equivalent if the momentum is assumed to be the product mq. Newton's formulation, however, is consistent with present-day relativity, whereas the other formulation has been disproved by the experiments on the deflection of fast electrons.

Newton's mechanics obtained its principal empirical confirmation from astronomical experience, because he could derive mathematically the planetary motions from its laws and from his law of gravitational attraction. But all the other great ideas of his work, the theory of tides, the calculation of the flattening of the earth and other planets, the derivation of the velocity of sound, etc., lent credence to the law of momentum, once they were verified by experience, because they are all based on it. In the following centuries, the thousandfold empirical verifications of Newton's mechanics produced an almost unlimited confidence in the law of momentum. In the minds of many scientists this law assumed the quality of a mathematical truth—which it is not.

The conservation of the angular momentum is one of the most important mathematical consequences of the law of linear momentum; the angular momentum is a directed quantity (a vector), which is determined by the momentum of a mass point and the radius vector assumed to be drawn to it from a fixed point in space. The angular momentum is perpendicular to these two vectors; its amount is obtained by multiplying the momentum component normal to the radius vector by the length of the latter. The earliest examples for the conservation of angular momentum are the two laws of Kepler, which state that the path of a planet is plane and that the radius vector drawn from the sun to the planet covers equal areas in equal times. Later on, the angular momentum law achieved prominence when, in 1765, Leonhard Euler (1707–1783), with analytical methods, developed the theory of the rotation of a rigid body about a fixed point; then, somewhat later (1834), Louis Poinsot (1777–1859) solved the same problem with the help of synthetic-geometrical methods. All that is needed in this case is the application and mathematical interpretation of the law of angular momentum. The top provides the most obvious example of the types of motion that can be treated with its help. When no point of the rigid body is held fixed, then the translatory motion of its center of mass is controlled by the law of linear momentum alone, and the rotation about this point by the law of angular momentum alone.

As a *physical* problem mechanics was completely solved by Newton. The mathematicians, however, worked on it for another century and a half. They erected a structure with an architectural beauty which Newton, himself, had never divined. The fact, however, that the architects were mathematicians and that they required no new experiment or new observations proves that the foundation laid by Newton was fully sufficient to support this structure. And, since the scientists of that time proposed to reduce all of physics to mechanics, they thought that it would also suffice to support the far greater building of physics. This belief inspired physical research till the end of the nineteenth century. Only then new ideas gained hold in mechanics as well as throughout the other branches of physics.

Newton's mechanics assumed action at a distance between the different bodies. His law of gravitational attraction shows this very distinctly. It is beside the point whether Newton himself had considered this law to be more than an approximation, to be replaced eventually by a law incorporating a finite velocity of propagation. The notion of the rigid body, where each force attacking one part affects the entire body instanteously, also is based on the idea of action at a distance. More fundamentally, the axiom of the equality of action and reaction shows the importance of action at a distance in Newton's theory: in case body A is the source of an action which changes its momentum and which *only later* reaches a body B, where it produces an equal and opposite change of momentum, the sum of the momenta of both bodies is evidently not the same during the interval of transfer as before and after. In a later section we are going to show how physics has been able to maintain the law of conservation of momentum by expanding the momentum concept.

We shall conclude this section with a remark concerning the invariability of mass. As mentioned above, this invariability was taken for granted because repeated weighings had never indicated any change in the weight of a body. After the revolutionary chemical discoveries of the eighteenth and nineteenth centuries, the question arose whether the chemical reactions left the total weight of the matter unchanged. Hans Landolt (1831–1914) achieved the highest accuracy for the necessary weight determinations for fifteen different reactions in a long series of tests which lasted from 1893 to 1909. He was able to exclude relative variations in weight greater than 10^{-6}. In a few cases he obtained even higher accuracy.

III. The Law of Energy

In mechanics, the beginnings of the knowledge of the law of energy coincide with those of the law of momentum, as shown by the aforementioned energy considerations of Galileo and Huygens. Newton's theory of planetary motion naturally includes the fact that the planet always possesses the same velocity when it returns to the same distance from the sun. For Newton, however, this was only one of the many conclusions from the law of momentum. Gottfried Wilhelm Leibniz (1646–1716) was the first to devote his attention to the product of mass and squared velocity, mq^2. In 1695 he called this expression *"Vis viva,"* and this term was used until the middle of the nineteenth century. Following Gustave Gaspard Coriolis (1792–1843), it was applied to one-half of mq^2, the value of $(m/2)q^2$. Johannes Bernoulli (1667–1748) gave us the expression *"energy,"* but this name did not gain the upper hand until later. We use "energy" today in a broader sense and denote the *"vis viva"* as *kinetic energy*. The kinetic energy first came to public attention through the well-known, endless quarrel between the Cartesians on the one hand and Leibniz and his disciples on the other, whether $m \cdot q$ or $m \cdot q^2$ represents the correct measure of force. Actually both quantities play important but different rôles in mechanics; the whole dispute concerned a fictitious problem, caused by the then ambiguous meaning of the word "force" (*vis*). In the eighteenth century, however, the force concept was still so overgrown with mysticism that profound thinkers were seriously concerned with the dispute.

A further step in the interpretation of the law of energy was taken by the above-mentioned Swiss mathematicians Johannes Bernoulli and Leonhard Euler. They emphasized that the change in kinetic energy in a closed mechanical system did not at all result in a reduction in its "capacity of action" (*Facultas agendi*), but only in its transition to other forms. In 1826 Jean Victor Poncelet (1788–1867) first introduced the term "work" for that product of a force by the path of its point of application, measured in the direction of the force, which in mechanics equals the change in energy. The kinetic energy is again of great importance in the papers on collision by Thomas Young (1773–1829), which were published in 1807. As one of the results of mechanics it was established by the end of the eighteenth century that it was impossible to construct with mechanical components a *perpetuum mobile,* a machine that would permanently create mechanical work from nothing. That this result had a more general significance was probably suspected even then; at least the Academy of Paris decided in 1775 not to accept any further supposed solutions of the problem of the *perpetuum mobile,* on the grounds that it had already wasted too much of its time examining these schemes. That these negative results could lead to important positive conclusions remained unrecognized at that time.

The stimulus for a generalization of the mechanical to a universal conservation law was furnished by the experience of long standing that kinetic energy or mechanical work could be lost while the temperature of the bodies involved was increased, as through friction, and also by the much more recently observed fact, on the steam engine, that one might gain work from thermic processes. It was with the steam engine that Sadi Carnot (1796–1832) concerned himself in 1824 in a very remarkable paper, which led to the basic result that the production of work was contingent on the transition of heat from the high temperature of the boiler to the lower temperature of the surroundings. His work was marred, however, by the erroneous belief, then current, that heat is an indestructible substance. His successor, Bénoit Clapeyron (1799–1864), still retained the same error in a paper in 1843. But even before, in the first decades of the nineteenth century, there had been voices which asserted the existence of a uniform "force," which was to account equally for the phenomena of heat, light, electricity, magnetism, chemical affinity, etc. Added to these were discussions of the metabolism of food in the body as the source of animal heat as well as their capacity to do work. A discovery was in the air, and several scientists independently made important contributions in its direction.

The first was Julius Robert Mayer (1814–1878), a physician, who (according to M. Planck's 1885 paper on the conservation of energy) "preferred, in his whole mental attitude, to generalize in the manner of a philosopher rather than build piecemeal and by experimental methods." In his brief paper of May, 1842, he

operated with the principles *"Ex nihilo nihil fit"* (from nothing
grows nothing) and *"Nil fit ad nihilum"* (nothing leads to nothing), applying them to the "power to fall," to motion, and to
heat; what is of permanent value in his paper is that he gives a
reasonably correct value for the mechanical equivalent of heat.
How he obtained his value, he does not tell until 1845; his computation is one still familiar, from the difference in the two
specific heat capacities of perfect gases, where he assumes by implication, but correctly, that their internal energy is independent
of the volume. Ludwig August Colding (1815–1888) obtained
almost precisely the same value in 1843 from experiments involving friction; however, his justification for the general conservation law appears to us even more fantastic than the one given
by Mayer. The latter considers electric and biological processes
already in his second publication, while in his third, in 1848,
he raises the question of the origin of solar heat, explains the
incandescence of meteors from their loss of kinetic energy in the
atmosphere, and applies the law of conservation to the theory
of tides. Quite obviously, Mayer fully realized the significance of
his discovery. Nevertheless, at first he remained quite unknown
and received the recognition due him only much later. Men like
Joule and Helmholtz cannot be blamed for not having known at
first of Mayer's sparsely distributed works and therefore for not
having quoted him.

Whatever we may think of Mayer's arguments, this much we
must admit: Since it is the task of physics to discover the general
laws of nature, and since one of the simplest forms of a general
law is to assert the conservation of some particular quantity, the
search for constant quantities is not only a legitimate line of
inquiry, but is exceedingly important. This approach was always
present in physics. We owe to this approach the early conviction
concerning the constancy of electric charge. The actual decision,
whether a quantity believed to be subject to a conservation law
does really possess that property can, of course, be reached only
by experimentation. The energy principle is an experimental
law just as much as the law of conservation of electricity. But
Mayer really took the road of the experiment in determining the
mechanical equivalent of heat. For other fields of physics, the
principle remained for him a program, to be carried through by
others.

As the second contributor we must mention James Prescott

Joule (1818–1889), who early in 1843 wrote a paper on the thermic and chemical effects of the electric current (this paper was not printed until 1846). He established by measurements that the heat developed in the electric wire of a galvanic cell (which later, appropriately, was called *Joule's heat*) equals the chemical heat of reaction if the reaction takes place in an open-circuited cell, and provided, as we must add today, that the current is produced by the cell without the development of heat. Shortly after and again in 1845, Joule reports determinations of the mechanical heat equivalent, in which he converts mechanical work into heat either directly, electrically, or through the compression of gases.

However, the man whose universally educated mind enabled him to develop the energy principle with all its implications was Hermann Ludwig Ferdinand von Helmholtz (1821–1894). Like Mayer, whose works he did not know and whose results he had to obtain independently, he approached the principle starting from medical investigations. In 1845 he had corrected, in a short paper, an error of Justus von Liebig's (1803–1873), by pointing out that one could not simply equate the heat of the combustion of food stuffs in the animal body to the heat of combustion of the constituent chemical elements; at the same time, he had made a brief survey of the implications of the energy principle for the different fields of physics.

His lecture before the Physical Society of Berlin, on July 27, 1847, goes farther into the ramifications of this idea. In contrast to Mayer, Helmholtz adopted the point of view of the mechanical nature of all processes of nature, to be comprehended by the assumption of attractive and repulsive central forces, as did most of his contemporaries. In this point of view he saw a sufficient and (erroneously) necessary condition for the impossibility of the *perpetuum mobile*. But in his deductions he did not use the mechanistic hypothesis at all, but derived the various expressions for the energy directly from the impossibility of the *perpetuum mobile*—if for no other reason than that the reduction of all processes to central forces had by no means been carried through. Thus, his results did not depend on the mechanistic hypothesis and therefore were able to survive it. His original contributions included the concept of potential energy (*"Spannkraft"*—stress) for mechanics, the expressions for the energy of gravitation, for the energy of static electric and magnetic fields, and his energy considerations applied to the production of electric current in galvanic cells and thermocouples as well as to electrodynamics, including electromagnetic induction. When, today, we compute the energy of a gravitational field as the product of masses and potentials, that of an electrostatic field as the product of charges and potentials, we are employing Helmholtz's methods directly.

It would lead us too far to go into the details. The further development of the principle cannot be treated here either. The final formulation only, due to William Thomson (later Lord Kelvin of Largs, 1824–1907) may be mentioned: "We denote as energy of a material system in a certain state the contributions of all effects (measured in mechanical units of work) produced outside the system when it passes in an arbitrary manner from its state to a reference state which has been defined *ad hoc*." The words "in an arbitrary manner" contain the physical law of the conservation of energy.

Helmholtz's considerations were not at all generally accepted; the older of his contemporaries were afraid of a revival of the phantasies of Hegel's natural philosophy, against which they had had to fight for such a long time. Only the mathematician Gustav Jakob Jacobi (1804–1851), who has made his own important contributions to mechanics, saw in them the logical continuation of the ideas of those mathematicians of the eighteenth century who had built up the science of mechanics. When, however, by about 1860 the law of energy had finally received general recognition, it became very soon a cornerstone of all natural science. Now every new theory, particularly in physics, was evaluated first by examining whether it was consistent with the law of energy. About 1890 many scientists, such as the well-known physical chemist Wilhelm Ostwald (1859–1932), were so enthralled by it that they not only undertook to deduce all other natural laws from it, but they actually made it the central thesis of a new *Weltanschauung,* energetics. Such exaggerations, however, were soon brought to an end by less excitable contemporaries.

The notion of energy also penetrated into engineering. Every machine was judged according to its balance of energy, i.e., the extent to which its energy input is transformed into the desired form of energy. Nowadays the energy concept is part of the working knowledge of every educated person. Only in regard to atomic physics, about 1924, was the principle of energy seriously doubted.

For a time Niels Bohr, H. A. Kramers, and J. C. Slater thought that energy was not conserved in the individual scattering process involving x-rays or γ-rays, but only as the average for many such processes. But in 1925 the famous coincidence experiments by Hans Geiger (1882–1945) and Walter Bothe established that their views were in error.

The problem of basing mechanics solely on the energy principle arose also in the epoch of energetics. We shall state the answer here, because it holds in modern relativistic mechanics even though originally it was obtained on the basis of Newton's mechanics. The law of momentum cannot be deduced from the principle of energy alone; it contains more than it. But if we add the principle of relativity—at that time of course only that of Galileo—to the law of energy, according to which there is not only one correct system of reference for the fundamental equations of mechanics but an infinite number of them, all of which move with respect to one another with constant velocity, in other words if we require that the sum of potential and kinetic energy is conserved in *each* system of reference of this kind, it necessarily follows that the total momentum of the closed system considered remains constant.[1] The connection between the two laws of conservation which is hereby revealed, is related to the inertia of energy, toward which our presentation is aimed.

But when the principle of energy was found to hold beyond mechanics, it first seemed to lose this connection entirely. It had to be accepted as being much more comprehensive than the purely mechanical law of momentum. How the law of momentum, too, gradually grew beyond the sphere of mechanics we are going to show in the following sections.

IV. The Theory of the Flow of Energy

Time changed, the knowledge of physics became more profound. No sharp dividing line, however, separates the epoch which concluded with the triumph of the principle of energy from the following period which is characterized by the displacement of the theories of action at a distance by the theories of local action, which better correspond to the principle of causality. The dates in our presentation clearly show that the transition was gradual and that the two epochs partly overlap.

The principle of local action and that of finite velocity of propagation, even in a vacuum, which is connected with it, first triumphed in electrodynamics; we are convinced of it, since Heinrich Hertz (1855–1894) discovered the electromagnetic waves in 1888. Nowadays we are also convinced that gravitation progresses with the speed of light. This conviction, however, does not stem from a new experiment or a new observation, it is a result solely of the theory of relativity. But that belongs to a later period.

Of course, the idea of local action for all domains filled with matter was known to former physicists. The oldest theory involving local action is contained in the theory of elastically deformable bodies, including fluid dynamics. The origins of that theory date back to the life-time of Newton. Not only was the *total* potential energy of a stressed body calculated without difficulty, but the potential energy could be localized, i.e., its share could be attributed to each part of the body. It was also generally accepted that a fluid moving under pressure not only conveys energy by transport, but also conducts an additional amount of energy, which is proportional to its velocity and pressure. In 1898 this theory was completed by Gustav Mie, when he taught us how to calculate the flow of energy for any motion of elastically stressed bodies. The assumption that energy flows like a substance can be carried through quite generally. In the driving belt, which connects the motor with a machine consuming energy, the energy flows against the motion of the stretched half of the belt. In the rotating and twisted shaft, which connects the engine of the ship with the propeller, it flows parallel to the shaft axis, i.e., at right angles to the velocity of the material parts.

For electromagnetic processes Helmholtz's method at first furnished merely a formula for the total energy; as long as one believed in action at a distance without a transmitting medium, the question of localization lacked meaning. But Michael Faraday (1791–1867), in the course of his long career of scientific investigation, developed the concept of the *field* as the medium transmitting such action; the field was considered as a change in the physical state of a system which was essentially located in the dielectric, or even in the empty space between the carriers of electric charge and electric current and between the magnets. With this approach, the problem of localization became significant; Maxwell's theory of electricity and magnetism, proposed in 1862, does, in fact, contain the expressions for calculating the energy density which is composed additively of an electric and a magnetic term. This development is a necessary supplement of the field concept.

J. H. Poynting (1852–1914) led the theory a step farther, by adding to it, in 1884, the notion of a *flux* of electromagnetic energy, long before Mie carried through this idea for the theory of elasticity. He took this step on the basis of a mathematical conclusion from Maxwell's equations, but accomplished much more, the creation of an entirely new physical concept. According to him, there is a flux of electromagnetic energy wherever an electric and a magnetic field are present at the same time. Now it was possible to determine the route by which the chemical energy, which in the galvanic cell is transformed into electromagnetic energy, gets to the wire that completes the circuit, where that energy, is converted into Joule's heat; likewise we can trace the energy on its way into the electric motor which transforms it into mechanical work. For us, this approach has become almost a matter of course; during Poynting's time, it was the cause of considerable conceptual difficulties and took a long time to gain acceptance.

For a special case the concept of a spatially distributed energy and its flow through empty space had already been developed, and this circumstance greatly facilitated the assimilation of Poynting's innovation. A body which radiates light or heat loses energy; this energy will not appear as the energy of a particular body until the radiation strikes another body. Therefore, if the sum total of all energies of the system is to remain constant, the radiated energy must in the meantime exist as *radiation energy*. And now Maxwell's theory, confirmed so brilliantly by Hertz's experiments, reveals light and heat radiation to be electromagnetic vibrations; its formulas for the density and for the flux of electromagnetic energy turned out to correspond precisely to the customary ideas concerning radiation. Thus physicists of the last decade of the nineteenth century gradually learned to appreciate the new concepts also for other electromagnetic fields.

But the concept of linear momentum required a similar generalization. Clerk Maxwell (1831–1879), in his comprehensive work, *Treatise of Electricity and Magnetism*, 1873, had shown that a body which absorbs a light ray experiences a force in the direction of the ray; its magnitude per unit cross section of the ray equals the energy flux S, divided by the velocity of light c. This assertion was confirmed experimentally much later, by P. Lebedew (1901), E. F. Nichols and G. F. Hull (1903), and, with an accuracy of about two parts in a hundred, by W. Gerlach and A. Golsen (1923). But even earlier there arose the question of the validity of the law of momentum.

Now it is true that the body which emits a ray experiences the opposite force of the one which absorbs it; and, since emission and absorption take equally long times, the two changes in linear momentum eventually compensate for each other exactly. However, while the ray passes from one body to the other, the total mechanical momentum is certainly different from that measured, either before the process of emission or after absorption. If we wish to maintain the law of conservation of momentum, we cannot but ascribe to the ray an *electromagnetic momentum* and to assert the law of conservation for the sum of mechanical and electromagnetic momenta. And then one cannot but extend this new concept to all electromagnetic fields, instead of restricting it to rapidly vibrating fields like heat and light radiations. It turns out that the field must contain momentum of the magnitude of S/c^2 per unit volume, where the symbol S denotes the magnitude of electromagnetic energy flux; both of these vectors possess the same direction.

$$g = S/c^2$$

is the momentum density of the field. Science owes this fundamental step to Henri Poincaré (1854–1912); his publication is contained in the Lorentz Anniversary Volume of 1900.

According to this theory, static fields, too, may give rise to energy flux, if only electric and magnetic fields overlap. The paths of this flow, though, are always closed, the energy current leads back into itself without being converted anywhere into other forms of energy. This conclusion has occasionally been used as a counter-argument in discussions of Poynting's law. As a matter of fact, in energy considerations it is perfectly permissible to disregard this closed energy current entirely.

But according to Poincaré, in such static fields we shall also encounter linear momentum. For the field as a whole, the total momentum is always zero; but the local momenta will in general give rise to an angular momentum whose sum total is different from zero. A system consisting of electric charges and of magnets at rest represents an electromagnetic top whose angular momentum corresponds to that of a mechanical top. As long as the state of the system remains unchanged, the angular momentum of either top remains unobservable. However, any change of the electromagnetic field, affecting either the magnitude or the direction of the angular momentum, must produce a torque on the material carriers of the electric charge and on the magnets, because the change in angular momentum of the field must be compensated for by a corresponding change of the mechanical angular momentum. This conclusion, which at first may appear surprising, actually results merely in the well-known forces experienced by moving charges in a magnetic field and by moving magnets in an electric field, once the calculations are carried out. But it is clear now that the electromagnetic momentum is observable not only in heat and light radiation.[2]

Of greater importance is another conclusion from the equation $g = S/c^2$. If we displace a carrier of electric charge, then the motion of the corresponding electric field gives rise to a magnetic field, and their coexistence leads both to a current of energy and to a momentum; if the system satisfies certain symmetry conditions, e.g., if the body is a sphere, these two vectors are parallel to the velocity of the body. Here, for the first time we encounter the *inertia* of electromagnetic energy; for this additional momentum represents an additional inertial mass. True, for a macroscopic body this additional mass is negligible for all charges that are experimentally feasible. But for the electron, considered a rigid sphere with a radius of about 10^{-13} cm, it was not difficult to find out that this additional inertial mass is of the same order of magnitude as the observed mass; in fact, it was suspected that the whole mass of the electron might be of electromagnetic origin. Many physicists tended toward this view soon after 1900. This hypothesis was examined most carefully by Max Abraham (1875–1922) in a famous investigation in 1903; the most striking result was that the electromagnetic momentum is proportional to the material velocity only for very small velocities, but otherwise increases more rapidly, in fact it increases without limit as the velocity of the electron approaches the velocity of light, c. The differentiation between two masses, both of which depend on the velocity, the *longitudinal mass* and the *transversal mass,* originated with this investigation; but it also showed that this result modifies the basic structure of mechanics only slightly. These investigations stimulated the performance of many experiments in which the deflection of fast electrons by electric and magnetic fields was measured with ever increasing accuracy. The results showed uniformly that the momentum increases more rapidly than the velocity, but even more rapidly than is predicted by Abraham's mechanics. These results led subsequently to a decision between his mechanics and relativistic mechanics.

For small velocities, both of the masses postulated by Abraham go over into the rest mass m, which is related to the electrostatic energy of the electron, E_0, by the equation

$$m = \frac{4}{3} \frac{E_0}{c^2}.$$

Theoretical investigations of the dynamics of cavity radiation inside a uniformly moving envelope are part of the same general approach. Since Planck's radiation law of 1900, the problem of cavity radiation at rest was completely settled. In 1904, F. Hasenöhrl (1874–1915) began to generalize the theory for cavities in motion, and K. v. Mosengeil and M. Planck completed the investigation with an improved theoretical approach. At rest all rays, no matter what their direction of propagation, have the same intensity, whereas in motion those are more intensive which form an acute angle to the direction of motion. As a consequence, we have a resultant momentum in the direction of motion. This momentum increases with the velocity, but more rapidly than the latter, and it would increase beyond all limits, if the velocity should approach that of light, c, but even more rapidly than in the case of Abraham's model of the electron. What is common to both cases is the relationship between the rest mass m and the rest energy E_0. We have here a special case of relativistic dynamics, but treated without explicit reference to the principle of relativity; the derivation from Maxwell's theory by itself assures consistency with that principle.

In the case of the sphere as well as in the case of cavity radiation, the total linear momentum is parallel to the velocity. In most other cases, however, the total momentum possesses a component at right angles to the velocity. Even when the motion is purely translatory, the angular momentum will change, and, to compensate for this change, the material carriers of the charge must experience a torque which will produce this compensatory change in the mechanical angular momentum. Thus it appeared reasonable that an electrically charged condenser, if suspended so that it could turn freely, would assume a particular orientation relative to the velocity of the Earth, the one in which that angular momentum vanishes. This conclusion is inescapable in Newtonian mechanics. However, in 1904 Fr. T. Noble and H. R. Trouton searched for this effect in vain, and even the much more accurate repetition of their experiment by R. Tomaschek (1925–26) showed no trace of the effect. Their result is just as convincing a proof of the principle of relativity as Michelson's interference experiment. Both of these experiments proved the necessity for a new mechanics; Michelson's experiment because it showed the contraction of moving bodies in the direction of motion, and the experiment by Trouton and Noble because it showed that an angular momentum does not necessarily lead to a rotation of the body involved.

Thus, a new epoch in physics created a new mechanics. This epoch, too, partly overlaps with the preceding one; it began, we might say, with the question as to what effect the motion of the Earth has on physical processes which take place on the Earth. Hendrik Anton Lorentz (1853–1928) had centered the general interest on this question through a famous essay in 1895. But in this case we can assign to the dividing line between epochs a precise date: It was on September 26, 1905, that Albert Einstein's investigation entitled "On the Electrodynamics of Bodies in Motion" appeared in the *Annalen der Physik*. Our presentation here need not report on the theory of relativity as a whole; we shall restrict ourselves to the question, how this theory led to the recognition that *all* forms of energy possess inertia.

And here we must first of all take exception to a prime example of national-socialist forgery of history, this time in the field of physics. Because of the work of Fritz Hasenöhrl which was mentioned above, Philipp Lenard and his cronies tried, during the period of the "Third Reich," to give credit for the law of inertia, which had received the limelight of attention because of its applications in nuclear physics, to this worthy physicist, who had long since died. But an examination of the literature shows conclusively that Hasenöhrl in 1904 applied the then current notion of the inertia of electromagnetic energy with partial success to the problem of cavity radiation. The idea of the inertia of other forms of energy occurred to him no more than to other physicists prior to Einstein.

Incidentally, we must emphasize that our division into epochs applies only to macrophysics. Simultaneously, the development of molecular physics proceeded at a different pace; initially it received more stimulus from macrophysics than it was able to return. Since Planck's law of radiation, which is based on molecular-statistical considerations, this relationship has undergone a gradual change, so that for later times we can no longer speak of two separate approaches in physics; but for our present purposes, this is rather unimporant.

V. The Inertia of Energy

The recognition of the inertia of energy as such was published by Einstein in 1905 (*Ann. d. Physik, 18,* 639), in the same year as his basic paper on the theory of relativity. Einstein derived this law relativistically. And, in fact, a rigorous derivation must start from there. But a year later Einstein gave another presentation which is merely approximate, but which possesses the great advantage of being more intuitive and of avoiding the relativistic foundation. For our purposes, it will be sufficient to trace the second argument.

Let a large cylindrical cavity which has been evacuated float in empty space; let its length be L and its mass M. Placed on its end faces there are two bodies, A and B, respectively, whose masses are sufficiently small compared to M that they can be disregarded in all sums involving M (Fig. 1). Let A transmit

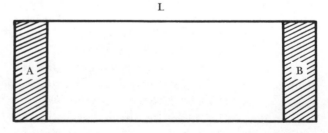

Fig. 1

to B an amount of energy ΔE in the form of Hertz vibrations or as light, and let B absorb that energy completely. The time required for emission and the equal time required for absorption shall be small compared to the time $T = L/c$, which is the time of travel through the interior of the cavity. During emission the body A receives a total impulse $G = \Delta E/c$, because of the radiation pressure, and through A, the cylinder as a whole receives the same impulse; the center of mass of the cylinder will therefore assume a velocity $q = G/M$ and, since it will retain this velocity during the time T, it will undergo a displacement

$$q T = \frac{L \Delta E}{M c^2}$$

in the direction $B \rightarrow A$.

After the body B has stored the energy received in some arbitrary form, we displace it by means of forces which originate inside the cylinder until it touches A. If we denote the mass of B at this stage by m_1, then the center of mass of the cylinder will be displaced by an amount $L m_1/M$ in the direction $A \rightarrow B$. Now let B transfer the energy ΔE back to A and subsequently return B to its original position, again by means of internal forces. If the mass of B is m_2 after it has lost the energy ΔE, then the displacement of the center of mass will amount to $L m_2/M$ in the direction $B \rightarrow A$. In the final state the distribution of energy is the same as it was initially: But there remains a resultant displacement of the center of mass of magnitude.

$$\frac{L}{M}\left(\frac{\Delta E}{c^2} + m_2 - m_1\right)$$

in the direction $B \rightarrow A$.

Shall we believe that the cylinder can shift its center of mass (and, therefore, in effect, itself) without any action from without and without any change in its interior? Such a possibility is not only inconsistent with mechanics but with our whole physical intuition, which, after all, contains a good deal of ancient valuable experience, even if that experience is often unconscious or uncomprehended. But the displacement will be zero only if the mass m_1 of the body possessing greater energy exceeds m_2, in fact if

$$m_1 - m_2 = \frac{\Delta E}{c^2}.$$

This is the amount by which the mass of a body must increase when it receives the energy ΔE, no matter in what form.

If q_1 was the velocity of B while it was moving toward A, then its momentum was

$$m_1 q_1 = \left(m_2 + \frac{\Delta E}{c^2} \right) \cdot q_1.$$

Let us divide this expression through by the volume V of the body B in order to obtain the momentum density (momentum per unit volume) and the energy density

$$\epsilon = \frac{\Delta E}{V},$$

to the extent that these quantities depend on the energy ΔE.

$$g = \frac{\epsilon}{c^2} q_1$$

is then the contribution of ΔE to the momentum density. On the other hand ϵq_1 is the flux S of the energy carried along convectively. As a result, we find the relationship

$$g = \frac{S}{c^2}$$

for the momentum density due to energy carried convectively.

Now we can modify the conceptual experiment which we have described by returning the energy ΔE to A not convectively, but for instance by means of a shaft which extends between A and B and which is rotated and twisted. We saw in Section IV that in such a shaft mechanical energy passes along the axis. Or, alternatively, we could return the energy by means of heat conduction. With every such type of energy flow there must be present an impulse in the same direction, if the center of mass of the cylinder is to return at the end to its original location. We require no lengthy calculations to show that momentum density and energy flux must always obey the same relationship indicated above.[3] Following Max Planck, who was the first to point out this relationship in 1908, we consider it the most general expression for the inertia of energy. The formula

$$m_1 - m_2 = \frac{E}{c^2}$$

makes use of the concept of mass, which loses its significance when momentum and velocity are no longer parallel to each other. And this is usually not the case, as was pointed out repeatedly; the rotating shaft is merely one striking instance.

Now if for a material body there is a difference in direction between its momentum and its velocity, then a purely translatory motion will change its angular momentum according to our previous remarks. Because of the law of conservation of angular momentum, it will require an angular momentum to *prevent* angular acceleration. This is the explanation for the experiment by Trouton and Noble. The condenser is stressed elastically by the forces of its own electric field. Its mechanical momentum therefore is inclined toward its velocity. The torque which, according to Section IV, the field exerts on the condenser is just sufficient to make possible its translatory motion.

But if an angular momentum is associated with translatory motion, is not the original law of inertia cancelled or at least its validity severely restricted? This law stated that a body *free of external forces* conserves its state of motion. Not at all! In the case of a force-body (an isolated static system, in technical langauge), the momentum is always in the direction of the velocity. Only for parts of the system can the momentum deviate from that direction. The material parts of the condenser are not free of force, they are subject to the action of the field. However, the physical system which consists of the condenser plates and the field is an isolated static system. Thus we must conclude that the condenser will not begin to rotate just because it participates in the translatory motion of the Earth.

Energy possesses inertial mass. Can, then, *all* of the inertial mass of a body be ascribed to its energy content? Do we have the farther reaching relationship $m = E/c^2$? The theory of relativity has given a positive answer to this question from the beginning. That is why it was able to combine the energy density, energy flux, and momentum density into a single mathematical quantity, the energy-momentum tensor. This development is due to Hermann Minkowski (1864–1909) who, in an exceedingly beautiful mathematical representation, introduced this quantity and rewrote the two conservation laws in the form of a single brief formula. Much more clearly than in Newtonian mechanics, we recognize here the fusion of the two laws with the help of a relativity principle. Experience has subsequently confirmed the validity of this daring step in a surprising manner. The last Section (VI) will deal with these developments.

First, however, we must explain why Abraham's model of the electron as well as cavity radiation yield the different relationship $m = (4/3)\ (E_0/c^2)$. The reason is the same in both cases. The electromagnetic field is not capable of existing by itself alone, it requires certain supports of a different nature ("material supports"). Cavity radiation can exist only within an envelope, and the charged sphere would fly apart if it were not for certain cohesive forces. In both cases, motion will give rise to an energy current within the material supports which is directed opposite to the motion. It contributes to the total momentum a negative amount and reduces the factor 4/3 to 1. (We are disregarding here the energy inherent in the supports themselves.) For the sphere, we have to consider in addition that, in contradiction to Abraham's assumptions, a moving sphere, because of its contraction in the direction of motion, turns into a rotational ellipsoid. That is why the dependence of the momentum on the velocity obeys relativistic dynamics rather than Abraham's.

The generalized law of the inertia of energy then provides that every inertial mass is due to energy, every momentum due to energy flux. The concept of mass formerly a basic concept of physics and a measure for the quantity of matter as such, is demoted to a secondary rôle. The law of conservation of mass is elminated; for the energy of a body can be changed by the transfer of heat or work. What is left of that law is absorbed by the energy law. On the other hand, the energy concept is expanded tremendously. We have reason to doubt that we know all forms of energy at the present time. But quite independently of that question, we can determine the total amount of energy in a body from its mass. We thereby get rid of the arbitrariness of the zero point of energy which the former definition of energy (cf. Section III) was forced to introduce. There are not merely energy *differences,* as before; the energy possesses a physically meaningful absolute value.

One type of energy, however, the new physics must eliminate from its list, and that is kinetic energy. For the energy E of a body possessing the velocity q, relativistic dynamics furnishes the equation

$$E = \frac{E_0}{\sqrt{1 - \left(\dfrac{q}{c}\right)^2}},$$

in which E_0 is the energy in the state of rest. Each type of energy therefore increases in the same manner as a result of motion. And in fact, if we consider all of the inertia as an attribute of energy, then we cannot base a particular type of energy in turn on the inertia. That is how fundamental the changes are which the law of the inertia of energy causes in our whole picture of the physical universe.

How do these conclusions agree with the conservation of mass as established by experiment, in particular with the accurate confirmations that Landolt achieved by his investigations of chemical reaction (cf. Section II)? After all, the reagents occasionally lose considerable amounts of heat in the course of the reaction.

Let us consider, as an instance, the combination of one mol of oxygen with two mols of hydrogen to form two mols of liquid water: if we compute from the well-known heat of reaction of this chemical reaction the corresponding loss of mass, we find that it amounts to less than the 10^{-10}th part of the total mass involved. Landolt's limit of accuracy was of the order of 10^{-6}. Thus he could not possibly have ascertained the relativistic loss of mass. For all changes of phase, such as the condensation from the gaseous state to a liquid or solid, or the solidification of a liquid, the mass losses are even much smaller.

However, already in his first paper, Einstein pointed out one possibility for determining the loss of mass connected with the transfer of energy: In radioactive nuclear transformations the amounts of energy liberated are much larger relative to the masses involved than even in the most powerful chemical reactions. The same is true of artificial nuclear transformations, about which we have learned so much since Lord Rutherford's great discovery in 1919. This aspect will be the topic of our last Section.

VI. The Inertia of Energy in Nuclear Physics

The chemist Jean Charles Galissard de Marignac of Geneva (1817–1894) in 1865 wrote in a paper concerned with completed and planned determinations of atomic weights in *Liebigs Annalen* as follows:

> If in these future determinations we should find to the same extent elements whose atomic weights approach integral values so remarkably close, then it appears inevitable to me that Prout's law be placed beside those of Gay-Lussac and Mariotte and that we recognize the presence of an essential cause according to which all atomic weights should exhibit simple numerical ratios, with secondary causes accounting for slight deviations from these ratios.

In all probability, Marignac merely expressed an opinion which was widespread at the time. Prout's hypothesis of the composition of all atoms from one or a few common original building stones has always been fascinating. Of course, science had to travel a long way before this opinion could assume definite shape. One of the preliminary conditions was, for instance, the discovery of isotopes by Frederic Soddy in 1910. Nowadays we know several atoms of different atomic weight at nearly each point of the periodic table; if we define the "unit of mass," by setting the atomic weight of the most frequent oxygen-isotope equal to 16.000 units, the atomic weight of each kind of atoms is really very close to an integer, the "mass number." This assertion, however, does not hold for the chemical elements composed of several isotopes. Another preliminary condition was the discovery of the neutron by J. Chadwick in 1930. Soon thereafter Jg.Tamm and Ivanienko, a little later W. Heisenberg, found that the composition of all atomic nuclei of protons and neutrons (which have nearly the same weight) is the essential cause for the integral numbers of the atomic weights, and that the energy loss accompanying the combination of these elementary particles and the loss of mass due to the inertia of energy is the secondary cause for the deviations from this law.

Two examples may show this: the atomic weights of the neutron and of the proton are 1.00895 and 1.00813 respectively;

their sum is 2.01708. The deuteron, however, which consists of a neutron and a proton, has only the atomic weight of 2.01472. The difference, amounting to 0.00235 units of mass, is the mass defect of the deuteron, the result of an energy loss of 3.51×10^{-6} erg due to the fusion of a neutron and a proton. The nucleus of a lithium atom consists of 3 neutrons and 3 protons; but this atom does not have the weight 3 times $2.01708 = 6.051124$, but only 6:01692; the mass defect, therefore, is 0.03432, corresponding to a loss of energy of 5.11×10^{-5} erg when this atomic nucleus is formed from its components. The mass defect increases with the atomic number up to 0.238 for uranium. The example of lithium shows that it can be as much as half a percent of the mass; this is quite a different order of magnitude from the above defect of 10^{-10} when water is formed.

The accuracy of the law of inertia, $E = mc^2$, was proved by W. Braunbeck (*Zs. f. Physik* 107, p. 1 (1937)), by calculating the velocity of light c according to that equation for a series of nuclear reactions for which the loss of energy and the masses of the reacting atoms before and after the reaction are measured independently of each other. For instance, he finds a mass defect of 0.02462 units of mass for the transmutation of the above-mentioned isotope of lithium with the mass number 6 plus a deuteron into two atoms of helium; this mass defect corresponds to 4.087×10^{-26} g; the loss of energy is 3.534×10^{-5}. This results in the velocity of light of 2.94×10^{-10} cm/sec. Forming the mean value over many different transmutations of nuclei, Braunbeck obtains the value 2.98×10^{10}, which is only 0.4 per cent below the value of 2.998×10^{10} cm/sec, measured electrically or optically.

All these tests, however, concern only the *variations* of energy and mass, not the entire mass of body. All the more important was a quite unexpected observation made in the Wilson chamber in connection with C. D. Anderson's discovery of the positron, the electron with a positive charge. In the proximity of an atomic nucleus, which serves only as a catalyst, γ-radiation can be transformed into a pair of electrons, a negative and a positive one. Not each γ-radiation can do this; its frequency ν has to be so high, its quantum of radiation $h\nu$ so great, that it is above 1.64×10^{-6} erg. As the rest mass of an electron—be it negative or positive—amounts to $9.1 \times 10^{-28}g$, it has a rest energy of

8.2×10^{-7} erg. This means: the γ-quantum has to be large enough to supply two electrons with their rest energy; what remains is used to give the electrons a certain velocity. The law of conservation of the electric charge is not disturbed here. The electron pair has the total charge zero.

Also, the reverse process of pair formation is possible. It is true, we have never observed in the Wilson chamber how a negative and a positive electron annihilate each other while emitting γ-radiation. But otherwise this annihilation radiation is well known experimentally; for example, Jesse DuMond measured its wavelength in 1949 with a crystal spectometer at 2.4×10^{-10} cm. According to all these observations, the law of inertia may be considered to be one of the results of physics which is best confirmed.

Thus we have followed the history of the laws of conservation of momentum and energy up to their amalgamation in the recent state of physics. Every thinking person cannot but be strongly impressed by the last consequence, which is extreme, but confirmed by experience, that at least for electrons the mass is nothing but a form of energy which can occasionally be changed into another form. Up to now our entire conception of the nature of matter depended on mass. Whatever has mass, —so we thought—, has individuality; hypothetically at least we can follow its fate throughout time. And this is certainly not true for electrons. Does it hold for other elementary particles, e.g., protons and neutrons? If not, what remains of the substantial nature of these elements of all atomic nuclei, i.e., of all matter? These are grave problems for the future of physics.

But there are more problems. Can the notions of momentum and energy be transferred into every physics of the future? The uncertainty-relation of W. Heisenberg according to which we cannot precisely determine location and momentum of a particle at the same time—a law of nature precludes this—, can, for every physicist who believes in the relation of cause and effect, only have the meaning that at least one of the two notions, location and momentum, is deficient for a description of the facts. Modern physics, however, does not yet know any substitute for them. Here we feel with particular intensity that physics is never completed, but that it approaches truth step by step, changing forever.

The Limits of Conservation

G. FEINBERG and M. GOLDHABER,
The Conservation Laws of Physics

In the course of the development of science, it was usually believed that those quantities which are conserved are among the basic concepts required to build a satisfactory physics. There have been many conservation principles which had the weight of the whole scientific community behind them, but which subsequently turned out to be failures. In the nineteenth century, the failure of conservation of heat, and in the twentieth the failure of conservation of mass are two notable examples.

The conservation laws, whether of linear momentum, energy, or charge, are empirical, and they may one day go the way of mass or heat. Hopefully, if they prove false, they will nevertheless become part of a more inclusive conservation law. But the evidence for the laws currently accepted as true ought to be systematically investigated so that the basis of belief in these principles is known.

In the following paper, G. Feinberg (1933–   ) and M. Goldhaber (1911–   ) not only carry through that task but they also bring out the subtlety and innovations in the way modern physics uses these laws. For example, the authors are not concerned primarily with whether conservation is true for all time. Modern physics has found that there are certain kinds of processes, strong and weak interactions, each taking a certain time (10^{-23} seconds for strong interactions and 10^{-10} seconds for the weak ones). These are regarded as the natural durations to consider, and the questions now asked are whether conservation is true for strong or weak interactions. If conservation fails, the law is not discarded. Even if a quantity is conserved for a very short time, but not for longer durations, the principle of conservation (suitably restricted) is still of good service for the investigation of the shorter, strong interactions.

The phenomenal accuracy of the laws of conservation of energy,

Reprinted with permission from *Scientific American*, vol. 209, no. 4, October 1963, pp. 36–45.

linear momentum, and charge is not only ground for awe. It is clear from the authors' work that such precision gives physicists confidence in the principle, and it tells them to what energy ranges they may have to go to find a breakdown. In this two-fold way, the discussion by Feinberg and Goldhaber underscores the empirical character of the laws of conservation.

The reader should also note that physicists today are not content merely with having such well-confirmed principles of conservation; they believe that these principles have their explanation in deeper principles that describe the symmetry properties of space, time, and other fundamental physical structures.

The philosopher Heraclitus taught that nothing is eternal; everything is continually in flux. There can be no doubt that over a period of time the objects around us do change in such respects as form and position. How these changes occur is described by certain of the laws of physics. The human mind, however, has always sought to find underlying the changes properties of physical objects that do not change. In the development of physics this search has been profoundly rewarded by the discovery of conservation laws. A conservation law states that the value of some quantity, such as energy or electric charge, does not change with time. The name suggests that although the form in which the quantity appears may vary, some essence is preserved.

The physicist's confidence in the conservation principles rests on long and thoroughgoing experience. The conservation of energy, of momentum and of electric charge have been found to hold, within the limits of accuracy of measurement, in every case that has been studied. An elaborate structure of physical theory has been built on these fundamental concepts, and its predictions have been confirmed without fail. Consequently there has been a tendency to forget that the basis of the conservation laws is, after all, empirical, and that there is always the possibility they may break down when they are extended into a new realm of physical phenomena.

Such a breakdown has come not once but several times in recent explorations of the elementary particles of matter. It turns out that the particles have some properties that are conserved only approximately; that is, the conservative "laws" for these properties hold on one time scale but break down on a longer time scale.

The "natural" period for interactions of elementary particles is about 10^{23} second. This is the time scale for the "strong" interactions that involve pions (pi mesons) and nucleons (protons or neutrons) and are responsible for the forces between particles in the nucleus of the atom. There is another class of particle processes called weak interactions, of which the best known is beta-decay (radioactive decay involving emission of an electron and a neutrino). The "weak" processes take at least 10^{10} second to occur, which is obviously much longer than the period of the strong interactions.

One conservation law that was found to break down was the conservation of "strangeness," a property associated with the so-called strange particles, such as the lambda, the sigma and the K. It was found that the total strangeness, obtained by adding the strangeness of the individual particles, does not change in strong interactions but need not be conserved in weak interaction [see "Elementary Particles," by Murray Gell-Mann and E. P. Rosenbaum; *Scientific American,* July, 1957].

This failure of a conservation law was disconcerting but did not surprise physicists very much, because the discovery of the exceptions came at the same time as the discovery of the law itself. Soon afterward, however, physicists were startled by the breakdown of a conservation law that had been honored for some time.

The event was the collapse of the conservation of parity, brought to light as the result of questions raised in 1956 by T. D. Lee and C. N. Yang. Again it was a case of a conservation law that had passed all its tests in one domain (strong interactions) but failed completely in another (weak interactions). This time physicists were stirred to profound questioning of some of their basic assumptions. Among other things, the parity failure served as a salutary reminder that no scientific belief, not even a well-established conservation law, can be taken for granted in areas where it has not been tested. The reminder has inspired many physicists to rigorous testing of the conservation laws, old and new, in the domain of particle physics. This article is a discussion of these recent tests.

In order to test a conservation law it is in principle necessary only to measure the conserved quantity at two different times. For several reasons the elementary particles lend themselves to more precise tests of conservation laws than do bodies of ordinary or of astronomical size. To begin with, the system to be studied must be isolated—free from any outside addition to or subtraction from the quantity whose conservation is being measured. It is not easy to isolate a large-scale body from all external influences; for example, the earth is bombarded by charged particles and is subject to the gravitational force of other celestial bodies. A system of elementary particles, on the other hand, can be isolated much more easily; it often isolates itself by the very rapidity of its interactions, because they take place much faster than any effect from its surroundings. Second, any violation of a conservation law is more likely to show up in the fundamental particles than in a large body, where the presence of a mixture of many atoms may mask slight discrepancies. Third, by making precise measurements of particle reactions it is often possible to say quantitatively how well we know that the conservation laws are satisfied.

Let us start with tests of the conservation of momentum— ordinary momentum in a straight line, not angular momentum. The most convenient experiment for this purpose is a collision (that is, an interaction) of two particles. The law of conservation of momentum states that the total momentum of the two particles will be precisely the same after the collision as it was before they collided.

Consider a head-on collision between two particles that are equal in mass and traveling at equal speeds. The momentum of each particle is its mass multiplied by its velocity. The two momenta are therefore equal, but since they have opposite directions the total momentum of the two particles is zero. If the conservation law holds, the total momentum after the collision must also be zero, and the two particles must rebound with equal and opposite momenta. They may fly off at angles to their original line of motion, but this does not matter as long as their momenta are equal and opposite. The collision may transform one or both of the particles; again conservation of momentum is not violated as long as the mass times velocity of one equals the mass times velocity of the other.

This is the content of the law of conservation of momentum. But the principle does not tell us anything about the speed of the individual particles after collision compared with its value before collision. For example, it would be possible for the original particles to come out moving much faster than they had come in without violating the law of conservation of momentum. But this does not happen because of an even more famous law: the law of conservation of energy.

In order to describe elementary-particle reactions in which the particle masses can change, it is necessary to use the conservation of energy in its modern form, which includes in the total energy the "rest-energy" of the particles. This rest-energy is given by the celebrated equation of Einstein $E = mc^2$: E is the rest-energy, m the rest-mass of the particle and c^2 the square of the speed of light. Different particles will in general have different rest-masses and hence different rest-energies. The total energy of a particle is the sum of this rest-energy and the kinetic energy of motion, which is a positive quantity also proportional to the rest-mass and which increases as the particle speed increases. For slow speeds the kinetic energy is proportional to the square of the particle's speed. As the speed approaches the speed of light, the mass increases and the formula for the energy is changed.

The law of conservation of energy then asserts that in any particle interaction the total energy of all the particles does not change. There are several predictions of the combined laws of energy and momentum conservation. If in a collision between particles of equal and opposite momentum the kind of particle is unchanged by the interaction, then the speeds should also be unchanged. If the particles are transformed into other particles with different masses, then the speeds of these emerging particles must differ from those of the original ones in a definite way.

Thousands of laboratory experiments, performed in different ways and measuring all the quantities involved, have confirmed that the laws of conservation of energy and of momentum do hold true in the domain of elementary particles. They have verified the formula that expresses the energy equivalent of the rest-mass. And some of the experiments have borne out the laws with a high degree of confidence in the accuracy of the results.

The situation is neatly illustrated by collisions between particles and their antiparticles. In such an encounter the particle and antiparticle annihilate each other and are transformed into photons or other particles that emerge with high energy. A meeting of an electron and its antiparticle the positron, for example, gives rise in less than 10^{10} second to two highly energetic photons, or gamma rays. When a proton and an antiproton meet, they produce a burst of pions and other particles.

In these annihilations the particle (electron or proton) is essentially at rest, and the antiparticle is often slowed almost to rest by electrical interaction of the atoms in the surrounding matter. Consequently the total momentum of the particle-antiparticle system is close to zero. The conservation of momentum then implies that the total momentum of the particles emerging must add up to zero. The rest-energy of the particle and antiparticle is converted into the total energy—kinetic energy plus rest-energy—of the products. A measurement of this energy should then agree with the original rest-energy.

In general the results of such experiments confirm the conservation of energy and momentum. But the accuracy of the measurements is limited. In the case of the electron-positron interaction the electron is never completely at rest, so that it is difficult to determine precisely if the electron and positron have a total momentum of zero when they collide. In the case of the proton-antiproton interaction the momenta of the pions emerging from the collision cannot be measured with an accuracy much better than one part in 1,000. All in all the accuracy of the measurements in the annihilation experiments is limited to the order of one part in 1,000 or one part in 10,000.

There is another type of experiment, however, in which the conservation of both energy and momentum has been checked to a far higher accuracy. This type of experiment has to do with the emission of gamma rays by "excited" nuclei. Is the total energy conserved when a nucleus emits a gamma ray? A very accurate means of determining this has been provided by the Mössbauer effect, for the discovery of which Rudolf Mössbauer was awarded a share of the 1961 Nobel prize in physics.

When a nucleus emits a gamma ray, the nucleus will normally recoil a bit to conserve momentum, just as a gun recoils when it fires a bullet. The total energy available must be shared between the photon and this recoil, so that the photon has a little less than the total energy. But suppose the nucleus is tightly bound in a crystal lattice, as it is in a metal. The inertia, or weight, of the crystal is so great compared with the photon's push that the bound nucleus does not recoil to any measurable degree. Hence the gamma ray gets essentially all the available energy and momentum.

Mössbauer pointed out that such a gamma ray, with unchanged energy, should readily be reabsorbed by a nucleus of the same kind as the one that emitted it. This "resonance absorption" turned out, indeed, to be an extremely sensitive and accurate measure of the energy of gamma rays. It is so accurate that by using the Mössbauer effect R. V. Pound and Glen A. Rebka of Harvard University were able to measure the very slight changes in a photon's energy that are produced by differences in the earth's gravitational field, thereby confirming a prediction of Einstein's general theory of relativity.

The Mössbauer effect itself is a striking confirmation of the conservation of energy. When the crystal-bound nucleus of an atom of the isotope iron 57, for example, emits a gamma ray photon, the photon can be absorbed by an iron-57 nucleus in another crystal. This means that the photon's frequency (that is, energy) must be in resonance with the frequency of the vibrations of the iron-57 nucleus, which in turn means that it must match the frequency with an accuracy of about one part in 10^{15}! That is a truly remarkable demonstration that the photon lost no energy and that energy is conserved in gamma ray emission. By the same token it also confirms the conservation of momentum; the special theory of relativity states that a photon, although weightless, has a momentum that is proportional to its energy (the momentum is equal to the particle's energy divided by the velocity of light). If the photon's energy remains unchanged, its momentum must be unchanged. It is clear that the laws of conservation of energy and momentum, introduced by Christian Huyghens, Daniel Bernoulli and Isaac Newton in the 17th century to describe collisions between macroscopic bodies, also apply with remarkable accuracy to the collisions and interactions of subatomic particles.

One useful aspect of the conservation formulas is that they enable us to tell which particle transformations or decays are possible and which are forbidden. The total energy of the products of an interaction or a decay must be equal to the rest-energy plus kinetic energy of the original particle or particles; it cannot be more and it cannot be less. It was this inviolable rule of energy conservation that led to the discovery of the neutrino and certain other previously unsuspected particles.

Energy and momentum are not the exclusive criteria of whether or not a given reaction can occur. For example, if we consider only the rest-mass of the electron, we can say that the electron could decay into a neutrino and a gamma ray without violating conservation of energy or momentum. But no such decay has ever been observed, and we believe it is actually forbidden by the law of conservation of electric charge.

This law was discovered in the 18th and 19th centuries by Benjamin Franklin and Michael Faraday. It was found then that electric charge occurs in two forms, positive and negative. The two kinds are never produced separately; they are always produced together and in equal amounts. Furthermore, the two cannot disappear separately, but equal amounts can combine to give neutral, or uncharged, matter.

In experiments with elementary particles a remarkable regularity of the electric charges has been found. If the charge is expressed in units in which the electron has one unit of negative charge and the positron one unit of positive charge, then all known particles either have one unit of positive or negative charge or are neutral. It is not known why there should be no particles whose charge differs from this unit, but the equality of charges has been verified to one part in 10^{17}.

When applied to particle reactions, the law of conservation of charge states that the total electric charge at the beginning is equal to the total electric charge at the end. To find the total charge the algebraic sum must be taken; that is, a positive charge counts as one unit, a negative charge as minus one unit and the two together come to zero total charge. It is now easy to see why the law forbids the transformation of an electron into a neutrino and a gamma ray. The last are uncharged; therefore the conversion would call for disappearance of the electron's unit of charge, which is forbidden. If it decays at all, the electron must decay into a charged particle.

In view of this, the law of conservation of energy tells us that in fact an electron cannot decay. There are no known charged particles with a smaller rest-mass, or rest-energy, than that of the electron. Unless and until such a particle is found, one must rule out any possibility of a breakdown of the electron, assuming that the laws of conservation of energy and of charge are valid.

Electric charge is conserved in all the particle reactions we know about. This does not necessarily verify, however, that it never disappears under any circumstances. We would like to have some upper limit for the rate of reactions that would not conserve the electric charge. One such reaction would be the decay of an electron; hence an experiment to test the conservation of charge could consist of watching and waiting for an electron to disappear. An experiment of this type was performed a few years ago at the Brookhaven National Laboratory.

The rationale of the experiment was as follows. Consider an electron in the filled, innermost shell (the K shell) of an atom. Suppose the electron were to "decay" in some way. One of the electrons in an outer shell would then drop into the hole left in the K shell and in doing so would release energy in the form of an X ray. In the experiment the electrons under surveillance were in iodine atoms in a crystal of sodium iodide. In this crystal the emergence of even a single X ray could be detected. And since the crystal contains a very large number of atoms, the counter monitoring it was in effect watching an immense population for just one electron decay occurring sometime, somewhere in that vast sea of electrons.

The experiment was set up and the crystal was monitored for several months. On the basis of the number of K electrons in the crystal it was calculated that an electron does not decay—that is, lose its charge—in at least 10^{17} years. This is 10^{24} times longer than the average time for a weak interaction of comparable energy, such as beta-decay. Therefore we can say that the law of conservation of charge is confirmed with an extremely high degree of probability.

There are still other conservation laws in the domain of particles. An example is the new law required to explain the stability of the particles in the nucleus. Consider the proton. None of the conservation laws we have discussed would prevent it from decaying into a positron and a gamma ray. In contrast to the electron, the proton is not the smallest known charged particle. There are several positively charged particles of smaller mass: the positron and various mesons. The proton could therefore break down without violating conservation of energy or of charge. Similarly, nothing in these conservation laws should forbid the neutron from decaying into a neutrino and one or more gamma rays.

Yet obviously these two particles, which make up the nuclei of atoms, do not decay, at least on any significant scale. If they did, the matter of our universe would have disintegrated into radiation long ago. Some property other than energy and charge must therefore be conserved. E. C. G. Stuckelberg and Eugene

P. Wigner have called the new conserved quantity "baryon number."

"Baryon" is the name assigned to the comparatively heavy particles: the proton and all other particles of the same or greater mass. Stuckelberg and Wigner have suggested that just as there is a quantum, or unit, of electric charge, so there is a quantum of the assumed baryon property. This quantum, or baryon number, is carried by the proton. The fact that the proton is the lightest particle carrying that quantity guarantees it against any decay. All heavier particles that can decay into a proton (for example the lambda particle, which decays into a proton with the emission of a pion) must have the same baryon number. Thus the baryon number is always conserved. The proton must have a positive baryon number and the antiproton a negative number of the same magnitude, so that the total baryon number of the two is zero. In their mutual annihilation there is no change of baryon number: it is still zero. As in the case of electric charge, in any reaction the total baryon number of all particles after the reaction must be the same as the total before.

How can one find limits within which the law of conservation of baryon number is known to be valid? The most critical test is similar to the one for the conservation of electric charge: watch the proton to see if it ever decays. If a proton decayed its breakdown would be easy to detect. It has so much more mass (that is, rest-energy) than any particle into which it might decay, such as a pion, that the conversion of the mass difference into energy would give the emerging particle or particles great kinetic energy. The signal of a proton breakdown would therefore be the appearance of a very energetic light particle in the collection of protons that is being watched.

A number of experiments monitoring large samples of material containing great numbers of protons have been carried out by Frederick Reines and Clyde L. Cowan, Jr., by one of us (Goldhaber), by Hans Frauenfelder and others. In none of these samples, each consisting of some 10^{30} protons, was proton decay detected. The experiments showed that a proton does not break down in at least 10^{22} years, which is 10^{43} times longer than the time of other particle decays of comparable energy.

The only problem with the law of conservation of baryon number is that the baryon number has no known properties other than that it satisfies the conservation law. It is not like electric charge, which can be defined and measured independently, for example by measuring the motion of charged particles in an electric field. The lack of such properties of the baryon number is disturbing, and the search for them is sure to continue. Meanwhile, in the absence of a firm theoretical basis for the law, we can only conclude that it is known to be true with great accuracy.

Are there other forms of conservation in the domain of elementary particles? Physicists have noted two that seem to be similar in kind (and in mysteriousness) to the conservation of baryon number. One applies to the leptons, the name given to the light particles (electrons, muons neutrinos and their antiparticles) to distinguish them from the baryons. It appears that the leptons possess a property called lepton number whose conservation forbids certain reactions the other conservation laws would allow (for example the transformation of a negative pion and a neutron into two electrons and a proton). The conservation of lepton number has been found to hold in all the cases so far studied.

The other new conservation law is connected with the recent discovery of the existence of two different neutrinos, one associated with muons and the other with electrons. Physicists now believe that the muon neutrinos, the muons themselves and their antiparticles have a property, called muon number, that forbids certain processes such as the decay of a muon into an electron and a gamma ray or the combination of a muon neutrino with a neutron to form a proton and an electron. The experimental absence of these reactions is the basis of the belief in the law of muon-number conservation.

Our purpose in this article has been to show the great power of the conservation laws and of the simple experiments that have been carried out to test them in the realm of elementary-particle processes. These laboratory tests have validated the laws to unprecedented limits of accuracy. They leave unanswered the question about the possibility that the conservation laws may not hold true on the cosmic scale, as the proponents of the theory of the continuous creation of matter in the universe contend. It must be said, however, that so far no one has demonstrated that it is necessary to assume such violations of the conservation laws do occur.

Finally, we should like to point out the usefulness of studying the conservation laws in the light of still more fundamental principles of physics. As an example, the law of conservation of momentum can be derived from the more basic concept that physical phenomena do not depend on the place where measurements are made. Such reductions of conservation laws to deeper principles may well lead to important clarifications of the still mysterious forms of conservation observed in the world of elementary particles.

The Principles of Symmetry

EUGENE P. WIGNER, *Symmetry and Conservation Laws*

It has been suggested (Feinberg and Goldhaber, p. 281) that current physics seeks to explain conservation laws by means of underlying principles of symmetry. This tendency is chiefly due to the pioneering work of E. P. Wigner (1902–   ). Wigner points out that there is a difference between the symmetry principles now studied, and certain earlier types. The earlier kind he calls geometrical, and the more recent type dynamical. The difference seems to be in the kind of entity whose symmetry is studied.

Let us suppose that a certain event e is followed by a series of events $e_1, e_2, \cdots$, which, taken together, have a certain structure. If we found that for any other event e', exactly like e, except for a difference in time, the structure of the ensuing events $e'_1, e'_2, \cdots$, was exactly like the original, then a temporal symmetry would have been discovered. If e^* differed from e only in its place of occurrence while the structure of events which followed was exactly like the original, then a spatial symmetry would have been discovered. These symmetries all have the following character: if any event e is related to e^* in a way R, then the structure of events following e is exactly like the structure of events following e^*. All such R symmetries are of the *geometrical* kind, according to Wigner's classification.

The second type of symmetry, *dynamical* symmetry, is not concerned with the structure of groups of events. Consider the make-believe situation of a certain physical quantity E (the electric field for example) being described with the aid of another kind of quantity A. It turns out, according to this imaginary example, that E can be described not only by the quantity A, but by A_1, A_2, and so forth. In fact if any two A's are related in a way G, then they both describe E, and conversely. The matter can be stated more briefly: a necessary and sufficient condition for two quantities to describe

Reprinted by permission from the *Proceedings of the National Academy of Sciences*, Vol. 51, 1964.

the same physical quantity E is that they be related to each other in the way G. Therefore, there is a wide choice in describing the quantity E. Suppose matters are now complicated by the discovery of a law which states how quantities like the A's change in time. After a time, the quantities A and A_1 change to A^* and A^*_1, respectively. If we chose two equivalent descriptions A and A_1, do the new quantities A^* and A^*_1 determined by the law continue to be equivalent descriptions? That is, are A^* *and* A^*_1 related in the way G? Dynamical symmetries seem to have the following character: if two quantities A and A_1 are related in the way G, and according to a law L the quantities are transformed into A^* and A^*_1, then the transformed quantities are also related to each other in the way G. The law L is then called *G-symmetrical*. The significant idea behind this kind of symmetry is that equivalent descriptions should remain equivalent even under the changes in the descriptions required by law.

The matter as we have described it is quite abstract, but one difference stands out between the geometrical and dynamical types of symmetry. The former states when two *events* are equivalent (as far as the structures of their ensuing events are concerned); the latter, when two *quantities* are equivalent for the description of a physical situation.

The search for dynamical symmetries is a very recent chapter in modern physics. Even so, Wigner seems to suggest that although there have been conservation principles which have already been explained by dynamical symmetries, this may not continue to be the case.

Introduction

Symmetry and invariance considerations, and even conservation laws, undoubtedly played an important role in the thinking of the early physicists, such as Galileo and Newton, and probably even before them. However, these considerations were not thought to be particularly important and were articulated only rarely. Newton's equations were not formulated in any special coordinate system and thus left all directions and all points in space equivalent. They were invariant under rotations and displacements, as we now say. The same applies to his gravitational law. There was little point in emphasizing this fact, and in conjuring up the possibility of laws of nature which show a lower symmetry. As to the conservation laws, the energy law was useful and was instinctively recognized in mechanics even before Galileo.[1] The momentum and angular momentum conservation theorems in their full generality were not very useful even though in the special case of central motion they give, of course, one of Kepler's laws. Most books on mechanics, written around the turn of the century and even later, do not mention the general theorem of the conservation of angular momentum.[2] It must have been known quite generally because those dealing with the three-body-problem, where it is useful, write it down as a matter of course. However, people did not pay very much attention to it.

This situation changed radically, as far as the invariance of the equations is concerned, principally as a result of Einstein's theories. Einstein articulated the postulates about the symmetry of space, that is, the equivalence of directions and of different points of space, eloquently.[3] He also re-established, in a modified form, the equivalence of coordinate systems in motion and at rest. As far as the conservation laws are concerned, their significance became evident when, as a result of the interest in Bohr's atomic model, the angular momentum conservation theorem became all-important. Having lived in those days, I know that there was universal confidence in that law as well as in the other conservation laws. There was much reason for this confidence because Hamel, as early as 1904, established the connection between the conservation laws and the fundamental symmetries of space and time.[4] Although his pioneering work was practically unknown, at least among physicists, the confidence in the conservation laws was as strong as if it had been known as a matter of course to all. This is yet another example of the greater strength of the physicist's intuition than of his knowledge.

Since the turn of the century, our attitude toward symmetries and conservation laws has turned nearly full circle. Few articles are written nowadays on basic questions of physics which do not refer to invariance postulates and the connection between conservation laws and invariance principles has been accepted, perhaps, too generally.[5] In addition, the concept of symmetry and invariance has been extended into a new area—an area where its roots are much less close to direct experience and observation than in the classical area of space-time symmetry. It may be useful, therefore, to discuss first the relation of phenomena, laws of nature, and invariance principles to each other. This relation is not quite the same for the classical invariance principles, which will be called geometrical, and the new ones, which will be called dynamical. Finally, I would like to review, from a more elementary point of view than customary, the relation between conservation laws and invariance principles.

Events, Laws of Nature, Invariance Principles

The problem of the relation of these concepts is not new; it has occupied people for a long time, first almost subconsciously. It may be of interest to review it in the light of our greater experience and, we hope, more mature understanding.

From a very abstract point of view, there is a great similarity between the relation of the laws of nature to the events on one hand, and the relation of symmetry principles to the laws of nature on the other. Let me begin with the former relation, that of the laws of nature to the events.

If we knew what the position of a planet is going to be at any given time, there would remain nothing for the laws of physics to tell us about the motion of that planet. This is true also more generally: if we had a complete knowledge of all events in the world, everywhere and at all times, there would be no use for the laws of physics, or, in fact, of any other science. What I am making is the rather obvious statement that the laws of the natural sciences are useful because, without them, we would know even less about the world. If we already knew the position of the planet at all times, the mathematical relation between these positions which the planetary laws furnish would not be useful but might still be interesting. It might give us a certain pleasure and perhaps amazement to contemplate, even if it would not furnish us new information. Perhaps also, if someone came who had some different information about the positions of that planet, we could more effectively contradict him if his statements about the positions did not conform with the planetary laws—assuming that we have confidence in the laws of nature which are embodied in the planetary law.

Let us turn now to the relation of symmetry or invariance principles to the laws of nature. If we know a law of nature, such as the equations of electrodynamics, the knowledge of the subtle properties of these equations does not add anything to the content of these equations. It may be interesting to note that the correlations between events which the equations predict are the same no matter whether the events are viewed by an observer at rest, or by an observer in uniform motion. However, all the correlations between events are already given by the equations themselves and the aforementioned observation of the invariance of the equations does not augment the number or change the character of the correlations.

More generally, if we knew all the laws of nature, or the ultimate law of nature, the invariance properties of these laws would not furnish us new information. They might give us a certain pleasure and perhaps amazement to contemplate, even though they would not furnish new information. Perhaps, also, if someone came around to propose a different law of nature, we could more effectively contradict him if his law of nature did not conform with our invariance principle—assuming that we have confidence in the invariance principle.

Evidently, the preceding discussion of the relation of the laws of nature to the events, and of the symmetry or invariance principles to the laws of nature is a very sketchy one. Many, many pages could be written about both. As far as I can see, the new aspects which would be dealt with in these pages would not destroy the similarity of the two relations—that is, the similarity between the relation of the laws of nature to the events, and the relation of the invariance principles to the laws of nature. They would, rather, support it and confirm the function of the invariance principles to provide a structure or coherence to the laws of nature just as the laws of nature provide a structure and coherence to the set of events.

Geometrical and Dynamical Principles of Invariance

What is the difference between the old and well-established geometrical principles of invariance, and the novel, dynamical ones? The geometrical principles of invariance, though they give a structure to the laws of nature, are formulated in terms of the events themselves. Thus, the time-displacement invariance, properly formulated, is: the correlations between events depend only on the time intervals between the events, not on the time at which the first event takes place. If P_1, P_2, P_3 are positions which the aforementioned planet can assume at times, t_1, t_2, t_3, it could assume these positions also at times $t_1 + t$, $t_2 + t$, $t_3 + t$ where t is quite arbitrary. On the other hand, the new, dynamical principles of invariance are formulated in terms of the laws of nature. They apply to specific types of interaction, rather than to any correlation between events. Thus, we say that the electromagnetic interaction is gauge invariant, referring to a specific law of nature which regulates the generation of the electromagnetic field by charges, and the influence of the electromagnetic field on the motion of the charges.

It follows that the dynamical types of invariance are based on the existence of specific types of interactions. We all remember having read that, a long time ago, it was hoped that all interactions could be derived from mechanical interactions. Some of us still remember that, early in this century, the electromagnetic interactions were considered to be the source of all others. It was necessary, then, to explain away the gravitational interaction, and in fact this could be done quite successfully. We now recognize four or five distinct types of interactions: the gravitational, the electromagnetic, one or two types of strong (that is, nuclear) interactions, and the weak interaction responsible for beta decay, the decay of the μ meson, and some similar phenomena. Thus, we have given up, at least temporarily, the hope of one single basic interaction. Furthermore, every interaction has a dynamical invariance group, such as the gauge group for the eletcromagnetic interaction.

This is, however, the extent of our knowledge. Otherwise, let us not forget, the problem of interactions is still a mystery. Utiyama[6] has stimulated a fruitful line of thinking how the interaction itself may be guessed once its group is known. However, we have no way of telling the group ahead of time, we have no way of telling how many groups and hence how many interactions there are. The groups seem to be quite disjointed and there seems to be no connection between the various groups which characterize the various interactions or between these groups and the geometrical symmetry group which is a single, well-defined group with which we have been familiar for many, many years.

Geometrical Principles of Invariance and Conservation Laws

Since it is good to stay on terra cognita as long as possible, let us first review the geometrical principles of invariance. This was recognized by Poincaré first, and I like to call it the Poincaré group.[7] Its true meaning and importance were brought out only by Einstein, in his special theory of relativity. The group contains, first, displacements in space and time. This means that the correlations between events are the same everywhere and at all times, that the laws of nature—the compendium of the correlations—are the same no matter when and where they are established. If this were not so, it would have been impossible for the human mind to find the laws of nature.

It is good to emphasize at this point the fact that the laws of nature, that is the correlations between events, are the entities to which the symmetry laws apply, not the events themselves. Naturally, the events vary from place to place. However, if one observes the positions of a thrown rock at three different times, one will find a relation between those positions, and this relation will be the same at all points of the earth.

The second symmetry is not at all as obvious as the first one: it postulates the equivalence of all directions. This principle could be recognized only when the influence of the earth's attraction was understood to be responsible for the difference between up and down. In other words, contrary to what was just said, the events between which the laws of nature establish correlations are not the three positions of the thrown rock, but the three positions of the rock, with respect to the earth.

The last symmetry—the independence of the laws of nature from the state of motion in which it is observed as long as this is uniform—is not at all obvious to the unpreoccupied mind.[8] One of its consequences is that the laws of nature determine not the velocity but the acceleration of a body: the velocity is different in coordinate systems moving with different speeds, the acceleration is the same as long as the motion of the coordinate systems is uniform with respect to each other. Hence, the principle of the equivalence of uniformly moving coordinate systems, and their equivalence with coordinate systems at rest, could not be established before

Newton's second law was understood; it was at once recognized then, by Newton himself. It fell temporarily into disrepute as a result of certain electromagnetic phenomena until Einstein re-established it in a somewhat modified form.

It has been mentioned already that the conservation laws for energy, linear and angular momentum are direct consequences of the symmetries just enumerated. This is most evident in quantum mechanical theory where they follow directly from the kinematics of the theory, without making use of any dynamical law, such as the Schrödinger equation. This will be demonstrated at once. The situation is much more complex in classical theory and, in fact, the simplest proof of the conservation laws in classical theory is based on the remark that classical theory is a limiting case of quantum theory. Hence, any equation valid in quantum theory, for any value of Planck's constant h, is valid also in the limit $h = 0$. Traces of this reasoning can be recognized also in the general considerations showing the connection between conserva-tion laws and space-time symmetry in classical theory. The con-servation laws can be derived also by elementary means, using the dynamical equation, that is, Newton's second law, and the assumption that the forces can be derived from a potential which depends only on the distances between the particles. Since the notion of a potential is not a very natural one, this is not the usual procedure. Mach, for instance, assumes that the force on any particle is a sum of forces, each due to another particle.[9] Such an assumption is implicit also in Newton's third law; otherwise the notion of counterforce would have no meaning. In addition, Mach assumes that the force depends only on the positions of the interacting pair, not on their velocities. Some such assumption is indeed necessary in classical theory.[5] Under the assumptions just mentioned, the conservation law for linear momentum fol-lows at once from Newton's third law and, conversely, this third law is also necessary for the conservation of linear momentum. All this was recognized already by Newton. For the conservation law of angular momentum which was, in its general form, dis-covered almost 60 years after the *Principia* by Euler, Bernoulli, and d'Arcy, the significance of the isotropy of space is evident. If the direction of the force between a pair of particles were not directed along the line from one particle to the other, it would not be invariant under rotations about that line. Hence, under the assumptions made, only central forces are possible. Since the torque of such forces vanishes if they are oppositely equal, the angular momentum law follows. It would not follow if the forces depended on the positions of three particles or more.

In quantum mechanics, as was mentioned before, the conservation laws follows already from the basic kinematical concepts. The point is simply that the states in quantum mechanics are vectors in an abstract space and the physical quantities, such as position, momentum, etc., are operators on these vectors. It then follows, for instance, from the rotational invariance that, given any state ϕ, there is another state ϕ_a which looks just like ϕ in the co-ordinate system that is obtained by a rotation a about the Z axis. Let us denote the operator which changes ϕ into ϕ_a by Z_a. Let us further denote the state into which ϕ goes over in the time interval τ by $H_\tau\phi$ (for a schematic picture, cf. Figure 1). Then,

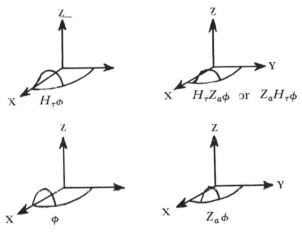

Fig. 1

because of the rotational invariance, ϕ_a will go over, in the same time interval, into the state $H_\tau\phi_a$ which looks, in the second co-ordinate system, just like $H_\tau\phi$. Hence, it can be obtained from $H_\tau\phi$ by the operation Z_a. It follows that

$$H_\tau Z_a\phi = Z_a H_\tau\phi \tag{1}$$

and since this is valid for any ϕ,

$$H_\tau Z_a = Z_a H_\tau. \tag{2}$$

Thus the operator Z_a commutes with H_τ, and this is the condition
for its being conserved. Actually, the angular momentum about
the Z axis is the limit of $(Z_a - 1)/_a$ for infinitely small a. The
other conservation laws are derived in the same way. The point is
that *the transformation operators, or at least the infinitesimal
ones among them, play a double role and are themselves the
conserved quantities.*

This will conclude the discussion of the geometrical principles
of invariance. You will note that reflections which give rise *inter
alia* to the concept of parity were not mentioned, nor did I speak
about the apparently much more general geometric principle of
invariance which forms the foundation of the general theory of
relativity. The reason for the former omission is that I will have
to consider the reflection operators anyway at the end of this discussion. The reason that I did not speak about the invariance
with respect to the general coordinate transformations of the
general theory of relativity is that I believe that the underlying
invariance is not geometric but dynamic. Let us consider, therefore, the dynamic principles of invariance.

Dynamic Principles of Invariance

When we deal with the dynamic principles of invariance, we
are largely on terra incognita. Nevertheless, since some of the attempts to develop these principles are both ingenious and successful, and since the subject is at the center of interest, I would like
to make a few comments. Let us begin with the case that is best
understood, the electromagnetic interaction.

In order to describe the interaction of charges with the electromagnetic field, one first introduces new quantities to describe the
electromagnetic field, the so-called electromagnetic potentials.
From these, the components of the electromagnetic field can be
easily calculated, but not conversely. Furthermore, the potentials
are not uniquely determined by the field, several potentials (those
differing by a gradient) give the same field. It follows that the
potentials cannot be measurable and, in fact, only such quantities
can be measurable which are invariant under the transformations
which transform a potential into an equivalent potential. This
invariance is, of course, an artificial one, similar to that which
we could obtain by introducing into our equations the locations
of a ghost. The equations then must be invariant with respect

to changes of the coordinate of that ghost. One does not see, in fact, what good the introduction of the coordinate of the ghost does.

So it is with the replacement of the fields by the potentials, as long as one leaves everything else unchanged. One postulates, however, and this is the decisive step, that in order to maintain the same situation, one has to couple a transformation of the matter field with every transition from a set of potentials to another one which gives the same electromagnetic field. The combination of these two transformations, one on the electromagnetic potentials, the other on the matter field, is called a gauge transformation. Since it leaves the physical situation unchanged, every equation must be invariant thereunder. This is not true, for instance, of the unchanged equations of motion, and they would have, if left unchanged, the absurd property that two situations which are completely equivalent at one time would develop, in the course of time, into two distinguishable situations. Hence, the equations of motion have to be modified, and this can be done most easily by a mathematical device called the modification of the Lagrangian. The simplest modification that restores the invariance gives the accepted equations of electrodynamics which are well in accord with all experience.

Let me state next, without giving all the details, that a similar procedure is possible with respect to the gravitational interaction. Actually, this has been hinted at already by Utiyama.[6] The unnecessary complication in this case is the introduction of generalized coordinates. The equations then have to be invariant with respect to all the coordinate transformations of the general theory of relativity. This would not change the content of the theory but would only amount to the introduction of a more flexible language in which there are several equivalent descriptions of the same physical situation. Next, however, one postulates that the matter field also transform as the metric field so that one has to modify the equations in order to preserve their invariance. The simplest modifications, or one of the simplest ones, leads to Einstein's equations.

The preceding interpretation of the invariance of the general theory of relativity does not interpret it as a geometrical invariance. That this should not be done, was pointed out already by the Russian physicist Fock.[10] With a slight oversimplification, one can say that a geometrical invariance postulates that two physically different situations, such as those on Figure 1, should develop in the course of time into situations which differ in the same way. This is not the case here: the postulate is merely that two different descriptions of the same situation should develop in the course of time into two descriptions which also describe the same physical situation. The similarity with the case of the electromagnetic potentials is obvious.

Unfortunately, the situation is by no means the same in the case of the other interactions. One knows very little about the weaker one of the strong interactions. The stronger one and also the weak interaction have groups which are first of all very much smaller than the gauge group or the group of general coordinate transformations.[11] Instead of the infinity of generators of the gauge and general transformation groups, they have only a finite number, that is, eight, generators. They do suffice, nevertheless, to determine the form of the interaction to a large extent, as well as to derive some theorems, similar to those of spectroscopy, which give approximate relations between reaction rates and between energies, that is, masses. Figure 2 shows the octuplet of heavy masses—

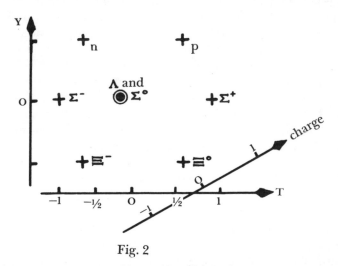

Fig. 2

its members are joined to each other by the simplest nontrivial representation of the underlying group which is equivalent to its conjugate complex.

Another difference between the invariance groups of electromagnetism and gravitation on one hand, and at least the invariance group of the strong interaction on the other hand, is that the operations of the former remain valid symmetry operations even if the existence of the other types of interactions is taken into account. The symmetry of the strong interaction, on the other hand, is "broken" by the other interactions; i.e., the operations of the group of the strong interaction are valid symmetry operations only if the other types of interactions can be disregarded. The symmetry group helps to determine the interaction operator in every case. However, whereas all interactions are invariant under the groups of the electromagnetic and gravitational interactions, only the strong interaction is invariant under the group of that interaction.

We have seen before that the operations of the geometric symmetry group entail conservation laws. The question naturally arises whether this is true also for the operations of the dynamic symmetry groups. Again, there seems to be a difference between the different dynamic invariance groups. It is common opinion that the conservation law for electric charge can be regarded as a consequence of gauge invariance; i.e., of the group of the electromagnetic interaction. On the other hand, one can only speculate about conservation laws which could be attributed to the dynamic group of general relativity. Again, it appears reasonable to assume that the conservation laws for baryons and leptons can be deduced by means of the groups of the strong and of the weak interactions.[12] If true, this would imply that the proper groups of these interactions have not yet been recognized. One can adduce two pieces of evidence for the last statement. First, so far, the conservation laws in question[13] could not be deduced from the symmetry properties of these interactions, and it is unlikely that they can be deduced from them.[14] Second, the symmetry properties in question are not rigorous but are broken by the other interactions. It is not clear how rigorous conservation laws could follow from approximate symmetries—and all evidence indicates that the baryon and lepton conservation laws are rigorous.[13] Again, we are reminded that our ideas on the dynamical principles of invariance are not nearly as firmly established as those on the geometrical ones.

Let me make a last remark on a principle which I would not hesitate to call symmetry principle and which forms a transition between the geometrical and dynamical principles. This is given by the crossing relations.[15] These relate to the probability amplitudes of any collision, such as

$$A + B + \cdots \to X + Y + \cdots \qquad (3)$$

This will be a function of the invariants which can be formed from the four-vector momenta of the incident and emitted particles. It then follows from one of the reflection principles which I did not discuss, the "time reversal invariance," that the amplitude of (3) determines also the amplitude of the inverse reaction

$$X + Y + \cdots \to A + B + \cdots \qquad (4)$$

in a very simple fashion. If one reverses all the velocities and also interchanges past and future (which is the definition of "time reversal"), (4) goes over into (3) so that the amplitudes for both are essentially equal. Similarly, if we denote the anti-particle of A by $\overline{A}$, that of B by $\overline{B}$, and so on, and consider the reaction

$$\overline{A} + \overline{B} + \cdots \to \overline{X} + \overline{Y} + \cdots \qquad (5)$$

its amplitude is immediately given by that of (1) because (according to the interpretation of Lee and Yang), the reaction (5) is obtained from (3) by space inversion. The amplitudes for

$$\overline{X} + \overline{Y} + \cdots \to \overline{A} + \overline{B} + \cdots \qquad (6)$$

can be obtained in a similar way. The relations between the amplitudes of reactions (3), (4), (5), (6) are consequences of geometric principles of invariance.

However, one can go further. The crossing relations tell us how to calculate, for instance, the amplitude of

$$\overline{X} + B + \cdots \to \overline{A} + Y + \cdots \qquad (7)$$

from the probability amplitudes of reaction (3). To be sure, the calculation, or its result is not simple any more. One has to consider the dependence of the reaction amplitude for (3) as an analytic function of the invariants formed from the momenta of the particles in (3), and extend this analytic function to such values of the variables which have no physical significance for the reaction (3) but which give the amplitude for (7). Evidently there are several other reactions the amplitudes of which can be obtained in a similar way; they are all obtained by the analytic continuation of the amplitude for (3), or any of the other reactions. Thus, rather than exchanging A and X to obtain (7), A and Y could be exchanged, and so on.

The crossing relations share two properties of the geometrical principles of invariance: they do not refer to any particular type of interaction and most of us believe that they have unlimited validity. On the other hand, though they can be formulated in terms of events, their formulation presupposes the establishment of a law of nature, namely the mathematical, in fact analytic, expression for the collision amplitude for one of the aforementioned reactions. One may hope that they will help to establish a link between the now disjoint geometrical and dynamical principles of invariance.

References

Natural and Forced Motions

ARISTOTLE, from *The Physics*

1. [This first paragraph ought to be attached to the last paragraph of the preceding chapter, which explained that while at any indivisible instant a moving object is "over against" a stationary object, so as to exactly "cover" it, it is not either moving or at rest "*at* that instant" (ἐν τῷ νῦν). This chapter should open with the review of Zeno's four arguments which follows.—C.]

2. At 233a21.—C.

3. The distance from the starting-point to the point at which the slower is overtaken by the swifter is easily calculated, if the start allowed and the respective velocities are given. This then is the definite line or distance (ἡ πεπερασμένη) which has to be covered, and if it is granted that the racers ever reach this point, it follows that the slower will be overtaken by the swifter. So that the problem is reduced to the question whether the racers can ever reach this particular point on their course. See Book VIII, chapter viii, 263a 4 *ff*.

4. [The translation here omits the parenthesis in the Greek: "(for the first C and the first B arrive at the opposite ends simultaneously)." This has already been stated above; it is repeated to justify the statement just made: that "the first B has passed all the C's *in the same time*" as the first C has passed all the B's (as stated in the previous sentence).—C.

5. The first C crosses all the B's while the first B crosses half the A's. Therefore while the first C crosses half the A's it will have time to cross all the B's (as it actually does, by the conditions of the problem). But it takes as long to pass an A as a B or C. Therefore half the time is as long as the whole time.

6. [I have printed Dr. Wicksteed's translation of this paragraph as it stands, with some verbal corrections, and given the text it implies. Mr. W. D. Ross has kindly allowed me to make use of an unpublished paper giving his interpretation of the Stadium and the readings he would adopt. I have made the following literal translation in accordance with his readings, adding a diagram which varies from the traditional one

309

only in placing the stationary A's outside the course in the position of the spectators.

'The fourth is the one about the equal bodies moving in the stadium past equal bodies in the opposite direction at equal speed, some (moving) from the end of the stadium, some from the *midway point (i.e.,* the turning-point in the double course). This, he thinks, involves the conclusion that the half-time is equal to *its double (the whole time).* The fallacy lies in the assumption that a body, moving with equal speed, takes an equal time in passing a moving body and a body of the same size that is at rest. This is false

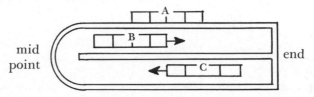

For instance, let the equal stationary bodies be $AA,$ and let $BB,$ *starting from the mid-point* [omit τῶν A, with EHI], be equal to them (the A's) in number and size, and let the CC *starting from the end (of the stadium)* be equal to them in number and size and equal to the B's in speed.

'Then it follows that the front B is opposite the (right-hand) end A (and the rear C) at the same time as the front C is opposite the (left-hand) end A (and the rear B), when they move past one another.

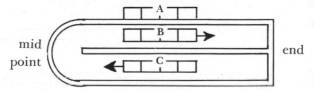

'And it follows that the (front) C has passed all the B's, while the (front) B has passed *half that number (of bodies* viz. the A's), so that the (B's) time is half (the C's time); for either takes the same time in passing each (body).

'And it follows that at the same moment the (front) B has passed all the C's; for the front C and the front B will arrive at the opposite end A's simultaneously.' [Omit ἴσον χρόνου . . . ὥς φησι: a gloss on l. 12 ἴσον γὰρ ἑκάτερόν ἐστι παρ' ἕκαστον because both the B's and the C's take the same time in passing the A's (παρὰ τὰ A).—C.]

7. It is more proper to say that during the change, it is 'always partially both but never either than to say that it is "always neither." '—Themistius.

8. See Book IV, chapter v., Introduction and text, also chapter iv., especially note a, Vol. I, p. 211.

9. Compare the definition of time in Book IV. 219b 3 ff.

10. [*Timaeus* 38 B. But Plato's own followers maintained that he was here using the language of "myth" and really regarded the visible universe and time as having no beginning.—C.]

11. [The reading is uncertain and the argument obscure. The previous argument about genesis (251a 16 ff.) was a dilemma. The corresponding dilemma here would be as follows: if there will ever be a time when there is no motion or change, then the things capable of causing or suffering change (τὰ κινητικὰ και κινητά) must either (a) perish after the last change has occurred or (b) endure unchanged for ever. Both suppositions should lead to an impossibility, viz. a change (destruction) occurring after what is *ex hypothesi* the last change.

The sentence οὐ γὰρ ἅμα παύεται κτλ seems to refer to supposition (a) and to mean that ceasing to be (or to exist as) a thing *capable* of causing or suffering change (τὸ παύεσθαι κινητικὸν ἢ κινητὸν ὄν) is a change that will have to follow ceasing to be a thing that is *actually* causing or suffering change (τὸ παύεσθαι κινοῦν ἢ κινούμενον) *i.e.*, the ceasing-to-be of potential agents and patients of change will be a change later than the last actual change they cause or suffer.

Does this sentence also prove the impossibility of the alternative supposition (b): that the potential agents and patients should endure unchanged for ever? Can it mean that the shift from being 'actually able' to cause and suffer change to perpetual immobility could only be affected by some change that rendered agents and patients permanently incapable of affecting one another—*e.g.*, put them out of range (*cf.* 251b 1 ff)—and this would be a change in their condition coming later than the last change? Then the ambiguous phrase τὸ παύεσθαι κινητικὸν ἢ κινητὸν ὄν would mean (not being simply blotted out, but) 'ceasing to be things actually capable of causing or suffering change.'

Unless we interpret so, alternative (b) seems to be ignored, for the next sentence, καὶ τὸ φθαρτικὸν, κτλ, cannot (with any ms. reading) be construed as referring to it. It appears to be an afterthought. If the κινητικά and κινητά are to be either destroyed or reduced to permanent incapacity, there must be something capable of destroying or immobilizing them (φθαρτικόν). 'And likewise that which is capable of destroying them will have to perish after they are destroyed, and what is capable of destroying *it* will have to perish in turn still later; for perishing is a kind of change' (and this change, or series of changes, will follow the last change).—C.]

12. For it is as good as saying there was a time when time was not. For the exact meaning of πλάσμα, *cf.* De Coelo 289a 6 and b 25.

13. [Or 'could only move with a single *kind* of motion,' whereas an animal can walk, run, leap, dance and move up or down (Themistius). As the text stands, this sentence seems to contain a separate objection, though the logic would be improved if the clause ὥστ' εἰ . . . τὸ κάτω (11.9-10) were transposed after ἑαυτὰ κινοῦσιν (1.11).—C.]

14. Cf. 227a 15.

15. [Or 'So none of these things (such as fire etc., which have a natural motion) moves itself—for they are of naturally coherent substance (cf. De gen. et corr. 327a 1, συμφυὲς ἕκαστον καὶ ἕν ὂν ἀπαθές)—nor yet does anything else that is continuous but the moving element in every case must be distinct from the moved, with such a distinction as can be actually seen in the case of an inanimate object (e.g., a boat) moved by an animate (a man rowing it).—C.]

16. A bar of iron is heavy, and so its inherent tendency is to pull or push things downwards. But it may be used as a lever and so made to impart a motion contrary to its own nature.

17. Because if you break the contact of the original magnet with the first bar of iron all the others instantly lose their power; whereas in the case of missiles each successive secondary agent becomes active as the previous one ceases to be so. So the power is not exhausted all at once when the first link in the chain ceases to exercise it, but is transmitted with gradually waning intensity to each successive link. [Cf. Plato, Ion 533 D, where Socrates tells Ion the rhapsode that when he recites there is a divine power of inspiration moving his hearers through him, 'like the power in the stone which Euripides calls the Magnesian stone though it is generally known as the stone of Heracles. This not only attracts rings that are made of iron, but puts into them the power of producing the same effect as the stone and attracting other rings in their turn. Sometimes there is quite a long chain of rings hanging from one another; but all the power they have depends on the stone.'—C.]

18. [Book III. chap. v., 205a 7 ff.—C.]

19. Aristotle himself does not accept this "general opinion." Only bodies have locality, and other things than bodies "exist." Cf. Introduction, pp. 268 ff, also Chap. iii. of this Book, on the ambiguity of "in."

20. On the Categories cf. Gen. Introd. pp. 1 sq.

21. ἐκ δὲ τῶν νοητῶν κτλ develops μέγεθος . . . ἔχει, while ἔστι δὲ τὰ μέυ κτλ develops σῶμα δ' οὐδέν.

22. Aristotle has already answered this by anticipation. The ultimate cosmic limits exercise an effective causation in actualizing the potentialities of buoyant and heavy bodies. Cf. 208 b 19.

23. [Diels, Vors. 19A 24.—C.]

24. We are to see later on that it is a fact that "heaven" is always moving (with the motion of rotation), but that it is a mistake (however natural) to infer that it must therefore be always changing its place, and so must have a place to change. Cf. chap. v.

25. [Prantl takes the clauses from ἐπεί (1. 24) to ἀπτομένων (1. 34) as a single sentence. The new point is in the apodosis, beginning ὅταν μὲν οὖν (1. 29).—C.]

26. [Or "and if anything that does not exist now, but formerly did exist, must have perished at some moment."—C.]

27. This is proved in Book VI. chap. i.

28. [Plato, according to Eudemus and Theophrastus (Simpl. 700. 18; Diels, *Dox*, 492).—C.]

29. [Pythagoreans, Diels, *Vors* 45B 33.—C.]

30. The characteristic motion of the heavens is re-entrant "circumlation." But part of a re-entrant circumlation is not re-entrant, whereas time (in addition to not being re-entrant, it might have been added, and therefore having no natural unit) is such that any portion of it (a day, for instance) is time just as truly and completely as any other portion (a day and night, for instance). So the Greek commentators understand this obscure and apparently not very significant passage. [If time is actually *identified* with the diurnal revolution, there is no time of less duration than a day.—C.]

31. If more or less change can take place in the same time, time itself cannot be identical with change. For if it were, equal portions of time would coincide with more or less of time itself.

32. [Sons of Herakles and of the daughters of Thespius, who were said to have colonized Sardinia (see Frazer on Apollodorus ii. 7. 6).—C.]

33. "From here to there" implies an interval between them, and this implies some kind of magnitude.

34. [Philop. 720. 26 paraphrases: "This before-and-after, when observed in motion and in time, in respect of the substrate (ὑποκείμενον)—that is the meaning of ὃ μέν ποτε ὄν—is nothing but the motion, but in respect of its definition is distinct: as ascent and descent in respect of their substrate are a ladder, but in definition distinct from it." Cf. *De. gen, et corr*. 319 b 2 ἥ ἔστι μὲν ὡς (sc ἡ ὕλη, the matter underlying the four simple bodies) ἡ αυτή, ἔστι δε ὡς ἑτέρα; ὁ μὲν γάρ ποτε ὄν ὑπόκειται, τὸ αὐτό, τὸ δ' εἶναι οὐ τὸ αὐτό "for that which underlies them, whatever its nature may be *qua* underlying them, is the same but its actual being is not the same" (Joachim).—C.] See pp. 209 and 213.

35. That is, "whose beginning and end are each a 'now.' "

36. The contrast is between the *numeri numerati* and the *numeri numerantes* and between the "concrete" and "abstract" of recent arithmetical terminology. In counting the successive 'nows,' we are counting sections of continuous time; but we are counting them in abstract numbers. Time, then, is a concrete numerable, not an abstract numerator.

Circular Inertia

GALILEO, from *Letters on Sunspots*

1. [The importance of this paragraph to the history of modern physics cannot be exaggerated. What it contains is the first announcement of the principle of inertia, according to which a body will preserve a state of uniform motion or of rest unless acted upon by some force.

Galileo's explicit statement of this principle is confined to the cases of (1) rotating bodies and (2) heavy bodies moving freely upon smooth spheres concentric with the earth. In applying the principle to physical problems, however, he included the more important case of bodies moving uniformly along straight lines, neglecting the force of gravitation. But even in such cases Galileo restricted his inertial principle to terrestrial objects. He did not, as is sometimes stated, attribute the orbital motions of the planets to an inertial principle acting circularly. In fact he did not attempt any explanation of the cause of planetary motions, except to imply that if the nature of gravity were known this too might be discovered (*Dialogue*, p. 235). The achievement of this prodigious step remained for Newton.—D.]

Motion: Uniform and Accelerated

GALILEO, from *Dialogue Concerning Two New Sciences*

1. ["Natural motion" of the author has here been translated into "free motion"—since this is the term used to-day to distinguish the "natural" from the "violent" motions of the Renaissance.—*Trans.*]

2. A theorem demonstrated on p. 175 below. [*Trans.*]

3. [As illustrating the greater elegance and brevity of modern analytical methods, one may obtain the result of Prop. II directly from the fundamental equation

$$s = 1/2g\,(t^2_2 - t^2_1) = g/2\,(t_2 + t_1)(t_2 - t_1)$$

where g is the acceleration of gravity and s the space traversed between the instants t_1 and t_2. If now $t_2 - t_1 = I$, say one second, then $s = (g/2)$ $(t_2 + t_1)$ where $t_2 + t_1$ must always be an odd number, seeing that it is the sum of two consecutive terms in the series of natural numbers. —*Trans.*]

Absolute and Relational Theories of Space and Time

LEIBNIZ and CLARKE, from *The Correspondence*

1. Despatched 25th Feb. 1716 (p. 193).

2. 'Purement'.

3. Transmitted 15th May 1716, delayed (p. 194).

4. This was occasioned by a passage in the private letter wherein Mr. Leibniz's third paper came inclosed. [Gerhardt says that there is no trace among the Leibniz papers of this letter. Klopp's edition of the Leibniz-Caroline correspondence also contains nothing relevant.]

5. Of nothing, there are no dimensions, no magnitudes, no quantity, no properties.

6. See above, §4 of my Second Reply.

7. Despatched with letter dated 2nd June 1716 (p. 195).

8. Sophia Electress of Hanover, mother of George I of England. Herren-hausen was the residence of the Electors of Hanover.

9. 'Il sera plus subsistant que les substances.'

10. Clarke's addition.

11. Epicurus held that while most atoms moved in regular courses, some occasionally made entirely uncaused swerves. Such a swerve occurring in the atoms of a man's brain gives rise to what the man regards as an act of free will.

12. Transmitted 26th June 1716 (p. 196).

13. Clarke qualifies the use of the term *property* in a note contained in the preface to Des Maiseaux's editions of the Correspondence. Cf. Introduction, p. xxix.

14. [Clarke quotes here from the General Scholium in the *principia*, the passage, 'He is eternal and infinite, . . . cannot be never and nowhere. He is omnipresent not virtually only but also substantially; for virtue cannot subsist without substance.' Appendix, pp. 167–8.]

15. Guericke (1602–86). Inventor of the air pump. He is said to have performed an experiment before the Emperor Ferdinand III in which he took two hollow copper hemispheres, exhausted the air from them with his pump, and then showed that thirty horses, fifteen pulling on each hemisphere, could not separate them. Leibniz corresponded with Guericke about the air pump in 1671–2 (G.I. 193).

16. Torricelli (1608–47). Pupil of Galileo, and inventor of the barometer. In his most famous experiment he took a long tube closed at one end, filled it with mercury and closing the open end with his finger, inverted it in a basin of mercury. When he removed his finger the level of mercury in the tube fell to thirty inches above the surface, leaving an apparent vacuum at the top of the tube.

17. 'au reste'.

18. 'au reste'.

On the Reality of Space
IMMANUEL KANT, from *Regions in Space*

1. [Euler there published his *Reflexions sur l'espace et le temps.*]

2. [Kant returned to this subject in 1786 in his treatise, *Was heisst: sich im Denken orientiren?*]

Time and Space as Conditions of Knowledge
IMMANUEL KANT, from the *Inaugural Dissertation*

1. [*catholica.*]

2. [*quae enim in sensus incurrunt.*]

3. [*immensi.*]

4. [*sed subjectiva condicio per naturam mentis humanae necessaria, quaelibet sensibilia certa lege sibi coordinandi.*]

5. [*relationes s. respectus.*]

6. What is simultaneous is not made simultaneous simply by not being successive. For when succession is taken away there is indeed removed a certain conjunction of things within the temporal series, but there does not on that account at once arise another real relation, such as is the conjunction of all in the same moment. For simultaneous things are joined in the same moment of time just as successive things in different moments. Thus though time possesses only one dimension, yet the ubiquity of time (to use Newton's manner of speaking) owing to which all things conceivable by sense are *at some time*, adds to the quantum of actuals a second dimension, so far as they hang, as it were, from the same point of time. For if you represent time by a straight line produced to infinity, and simultaneous things at any point of time by lines drawn perpendicular to it, the plane thus generated will represent the phenomenal world, both as to its substance and as to its accidents.

7. [*cerni.*]

8. I here pass over the proposition that space must necessarily be conceived as a continuous quantum, since that is easily demonstrated. Owing to this continuity, it follows that the simple in space is not a part but a limit. But a limit in general is that in a continuous quantum which contains the ground of its limits [*limes;* elsewhere, throughout, the term used is *terminus*]. A space which is not the limit of a second space is complete, i.e., a solid. The limit of a solid is a surface, of a surface a line, of a line a point. Thus there are three sorts of limits in space, just as there are three dimensions. Of these limits two, surface and line, are themselves spaces. The concept of limit has no application to quanta other than space and time.

9. [*sub ulla specie.*]

10. [*a sensu objectorum.*]

11. [*perpetuas.*]

On the Law of Inertia

ERNST MACH, from *The Science of Mechanics*

1. On the physiological nature of the sensations of time and space. Cf. *Analyse der Empfindungen*, 6th ed. [English ed. *The Analysis of Sensations* (1914)]; *Erkenntnis und Irrtum*, 2nd ed.

2. *Principia*, p. 19, Coroll. V: "The motions of bodies included in a given space are the same among themselves, whether that space is at rest or moves uniformly forwards in a right line without any circular

motion." [See the article on "Absolute Space, Time and the Science of Motion."—Koslow]

3. *Ibid.*

The Definition of Mass

ERNST MACH, from *History and Root of the Principle of the Conservation of Energy*

1. [Foucault, French physicist (1819–1868), demonstrated the axial rotation of the Earth by the apparent clockwise motion of a pendulum's plane of oscillation (1851).—Koslow]

2. *Ueber die Definition der Masse, Repertorium für physikalische Technik* . . . . , Bd. IV, 1868, pp. 355 sqq. [Prague, November 15, 1867.]

3. [Refers to Newton's "Law III" principle stated on p. 81.—Koslow]

The Simultaneity of Events

ALBERT EINSTEIN, from *On the Electrodynamics of Moving Bodies*

1. Translated from "Zur Elecktrodynamik bewegter Körper," *Annalen der Physik,* 17, 1905.

2. The preceding memoir by Lorentz was not at this time known to the author. [Lorentz, "Electromagnetic phenomena in a system moving with any velocity less than that of light," *Proc. of the Acad. Sci. Amsterdam,* 6, 1904.]

3. i.e., to the first approximation.

4. We shall not here discuss the inexactitude which lurks in the concept of simultaneity of two events at approximately the same place, which can only be removed by an abstraction.

5. "Time" here denotes "time of the stationary system" and also "position of hands of the moving clock situated at the place under discussion."

6. [We omit here, the derivation of the equations.—Koslow]

7. That is, a body possessing spherical form when examined at rest.

8. Not a pendulum-clock, which is physically a system to which the Earth belongs. This case had to be excluded.

Relativity and the Problem of Space

ALBERT EINSTEIN, from *Relativity, the Special and the General Theory*

1. This expression is to be taken *cum grano salis.*

2. Kant's attempt to remove the embarrassment by denial of the ob-

jectivity of space can, however, hardly be taken seriously. The possibilities of packing inherent in the inside space of a box are objective in the same sense as the box itself, and as the objects which can be packed inside it.

3. For example, the order of experiences in time obtained by acoustical means can differ from the temporal order gained visually, so that one cannot simply identify the time sequence of events with the time sequence of experiences.

4. This inexact mode of expression will perhaps suffice here.

5. If we consider that which fills space (e.g., the field) to be removed, there still remains the metric space in accordance with (1) $[ds^2 = dx_1^2 + dx_2^2 + dx_3^2 - dx_4^2]$, which would also determine the inertial behavior of a test body introduced into it.

The Origin of Inertia

D. W. SCIAMA, from *The Unity of the Universe*

1. Except that like charges *repel* one another.

2. For simplicity, I am leaving out a numerical factor which does not differ much from unity.

3. And, of course, to the inverse square of their distance apart.

4. The forces involve the vector product of the velocity and the magnetic field in one case, and the velocity and angular velocity of the rotating frame in the other.

The Mechanical Equivalent of Heat

J. P. JOULE, from *Heat and the Constitution of Elastic Fluids*

1. Read at a Meeting of the Manchester Literary and Philosophical Society, October 3, 1848, and published in the Society's Memoirs, November 1851.

2. Phil. Mag. vol. xxiii, pp. 263, 347, 435.

3. The equivalent I have since arrived at is 772 foot-pounds. [See Phil. Trans. 1850, part 1.—May 1851, J.P.J.]

4. Phil. Mag. vol. xxvi.

5. I subsequently found that M. Mayer had previously advocated a similar hypothesis, without, however, attempting an experimental demonstration of its accuracy. [—*Annalen* of Wöhler and Liebig for 1842.—May 1851, J.P.J.]

6. A complete theory of the motive power of heat has been recently communicated by Professor Thomson to the Royal Society of Edinburgh. In this paper the very important law is established, that the

fraction of heat converted into power in any perfect engine, is equal to the range of temperature divided by the highest temperature above absolute zero. [May 1851, J.P.J.]

7. *Elements of Chemical Philosophy*, p. 95.

8. Mr. Rankine has given a complete mathematical investigation of the action of vortices, in his paper on the Mechanical Action of Gases and Vapours. [—Trans. Royal Society of Edinburgh, vol. xx. part I.—May 1851, J.P.J.]

9. If we assume that the particles of gas are resisted uniformly until their motion is stopped, and that then their motion is renewed in the opposite direction, by the continued operation of the same cause, as in the projection upwards and subsequent fall of a heavy body; the maximum velocity of the particles will be to the uniform velocity required by the theory assumed in the text, as the square root of two is to one, and the comparison of the theoretical with the experimental specific heat will be as follows:—

	Experimental specific heat	Theoretical specific heat
Hydrogen	2·352	3·012
Oxygen	0·168	0·188
Nitrogen	0·195	0·214
Carbon oxide	0·158	0·136

I have just learned that the experiments of Regnault on the specific heat of elastic fluids are on the eve of publication, and doubt not that their accuracy will enable us to arrive at a decisive conclusion as to the correctness of the above hypothesis. [—June 1851, J.P.J.]

Conservation of Charge

MICHAEL FARADAY, from *On Static Electrical Inductive Action*

1. See Experimental Researches, Par. 1295, &c., 1667, &c., and "Answer to Dr. Hare," Philosophical Magazine, 1840, S.3. vol. xvii, p. 56, viii.

Conservation and the Meaning of Charge

MAXWELL, from *A Treatise on Electricity and Magnetism*

1. [The electrification of a body refers to the total charge on it.—Ed.]

2. This, and several experiments which follow, are due to Faraday, 'On Static Electrical Inductive Action,' *Phil. Mag.*, 1843, or *Exp. Res.*, vol. ii, p. 279.

3. [This is an illustration of Art. 100c.]

4. [The difficulties which would have to be overcome to make several of the preceding experiments conclusive are so great as to be almost insurmountable. Their description however serves to illustrate the properties of Electricity in a very striking way. In Experiment V no method is given by which the electrification of the outer vessel can be measured.]

5. 'On static Electrical Inductive Action,' *Phil. Mag.*, 1843 or *Exp. Res.*, vol. ii.

6. [See Maxwell's "Letter to P. G. Tait," p. 245.—Koslow]

On the Interaction of Natural Forces

HERMANN HELMHOLTZ, from *Popular Scientific Lectures*

1. [See Leibniz: "On the Quantity of Motion," p. 225.—Koslow]

Conservation of Mass

ALBERT EINSTEIN, from $E = MC^2$

1. [See Joule, p. 204.—Koslow]

The Principles of Conservation

MAX VON LAUE, *Inertia and Energy*

1. Let us consider an isolated system of any number of mass-points m_i with the velocities q_i relative to a first system of reference, the potential energy of which is a function of the relative co-ordinates of the m_i such as Φ. From the law of energy we have:

$$1/2 \; \Sigma \; m_i q_i^2 + \Phi = C,$$

where C is a quantity independent of time.

In a second system of reference which moves with the constant velocity v relative to the first system, the mass-point m_i has the velocity $q_i - v$. Relative to this system the law of energy becomes:

$$1/2 \; \Sigma \; m_i \; (q_i - v)^2 + \Phi = C'$$

Again C' is independent of time; Φ has the same value in both cases. Subtracting the first equation from the second we therefore obtain:

$$1/2 \; v^2 \; \Sigma \; m_i - (v \cdot \Sigma \; m_i q_i) = C' - C.$$

According to this equation the scalar product $(v \cdot \Sigma \; m_i q_i)$ is independent of time, no matter what direction and amount v may have. And this is possible only if $\Sigma \; m_i q_i$, the total momentum of all mass-points, is constant with time.

2. If we permit ourselves in passing to operate with a (fictitious) single magnetic pole of magnitude m and if we form a system containing in

addition a single electric charge of magnitude e, then the angular momentum of the field has the direction from e toward m, if both are positive or both negative (otherwise the direction is reversed), and its magnitude is independent of the distance and equals em/c. Starting with this law, and by applying vector addition, we can get the angular momentum for a charge in the field of a magnetic dipole and also for more complicated cases.

3. If S is the density of the energy current in the shaft or in the heat conductor that returns the energy ΔE to A, if Q is the cross section of the carrier of this energy current and T' the time that it takes the current to return the energy ΔE, then we find

$$SQT' = \Delta E.$$

On the other hand, if G is the momentum possessed by the shaft or heat conductor because of the energy current in the direction $B \rightarrow A$ and g the momentum density, then

$$G = gQL,$$

and

$$G/M = gQL/M$$

is the velocity which the center of mass of the cylinder receives because it must acquire the compensatory momentum $-G$; the direction of the velocity is from A toward B. During the time T', the center of mass will travel a distance

$$G'\,T'\,/M = GQLT'/M.$$

Now this displacement must equal the preceding displacement in the direction $B \rightarrow A$.

$$L \cdot \Delta E/Mc^2 = SQLT'/Mc^2$$

and we conclude again that

$$g = S/c^2.$$

In this connection, one should not object that in the event of heat conduction or other irreversible processes the entropy will increase and that at least in this respect the initial condition of the system cannot be recovered. Increase in entropy by itself cannot cause a displacement of the center of mass, if for no other reason than that the direction of the displacement would remain completely undetermined.

The Principles of Symmetry

EUGENE P. WIGNER, *Symmetry and Conservation Laws*

1. G. Hamel in his *Theoretische Mechanik* (B. G. Teubner, 1912) mentions (page 130) Jordanus de Nemore ($\sim$1300) as having recognized essential features of what we now call mechanical energy and Leonardo da Vinci as having postulated the impossibility of the perpetuum mobile.

2. F. Cajori's *History of Physics* (New York: Macmillan Company, 1929) gives exactly half a line to it (page 108).

3. See, for instance, his semipopular booklet *Relativitätstheorie* (Braunschweig, Friedrich Vieweg und Sohn, various editions, 1916–1956).

4. Hamel, G. Z. *Math. Phys.*, vol. 50:1 (1904); also Engel, F., *Abhandl. Ges. Wissensch. Göttingen, Math-Physik. KL.,* 270 (1916).

5. See the present writer's article in *Progr. Theoret. Phys.* vol. 11:437 (1954); and more recently, Greenberger, D.M., *Ann. Phys.*, vol. 25:290 (1963).

6. Utiyama, R., *Phys. Rev.*, vol. 101:1597 (1956); also Yang, C. N., and Mills, R. L., *Phys. Rev.*, vol. 96:191 (1964)

7. Poincaré, H., *Compt. rend. Soc. biol.*, vol. 140:1504 (1905); *Rend. Circ. Math. Palermo,* vol. 21:129 (1906).

8. Thus Aristotle's *Physics* postulated that motion necessarily required the continued operation of a cause. Hence, all bodies would come to an absolute rest if they were removed from the cause which imparts them a velocity (cf. e.g. Crombie, A. C., *Augustine to Galileo,* London, Falcon Press, 1952, page 82 or 244). This cannot be true for coordinate systems moving with respect to each other. The coordinate systems with respect to which it is true then have a preferred state of motion.

9. Mach, E., *The Science of Mechanics* (various editions, Chicago, Open Court Publishing Company), chapter III, section 3.

10. Fock, V., *The Theory of Space, Time and Gravitation,* New York, Pergamon Press (1959). Similar ideas were proposed before by Kretschman, E., *Ann. Physik.*, vol. 55:241 (1918).

11. For the strong interaction, compare Ne'eman, Y., *Nucl. Phys.*, vol. 26:222 (1961), and Gell-Mann, *Phys. Rev.*, vol. 125:1067 (1962). For the weak interaction, Feynman, R. P., and Gell-Mann, M., *Phys. Rev.*, vol. 109:193 (1958) and Sudarshan, E. C. G., and Marshak, R. E., *Phys. Rev.*, vol. 109:1960 (1958); also Sakurai, J. J., *Nuovo Cimento,* vol. 7:649 (1958), and Gershtein, G. S., and Zeldovitch, A. B., *J. Exp. Theor. Phys. USSR,* vol. 29:698 (1955).

12. For the baryon conservation law and the strong interaction, this was suggested by the present writer, *Proc. Am. Phil. Soc.,* vol. 93:521 (1949, and *Proc. Nat. Acad. Sc.,* vol. 38:449 (1952). The baryon conservation law as first postulated by Stueckelberg, E. C. G., *Helv. Phys. Act,* vol. 11:299 (1938).

13. For the experimental verification of these and the other conservation laws, see Feinberg, G. and Goldhaber, M., *Proc. Nat. Acad. Sc.,* vol. 45:1301 (1959). The conservation law for leptons was proposed by Marx, G., in *Acta Phys. Hung.,* 3:55 (1953); also Zeldovitch, A. B., *Dok. Akad. Nauk USSR,* vol. 91:1317 (1953), and Konopinski, L. J. and Mahmoud, H. M., *Phys. Rev.,* vol. 92:1045 (1953). It seemed to be definitely established by Lee, T. D., and Yang, C. N., *Phys. Rev.* vol. 105:1671 (1957).

14. For the baryon conservation and strong interaction, this was emphatically pointed out in a very interesting article by Sakurai, J. J., *Ann. Phys.,* vol. 11:1 (1960). Concerning the conservation of lepton member, see Marx, G. Z., *Naturforsch,* vol. 9a:1051 (1954).

15. Goldberger, M. L., *Phys. Rev.,* vol. 99:979 (1955); Gell-Mann, M., and Goldberger, M. L., *Phys. Rev.,* vol. 96:1433 (1954) See also Goldberger, M. L., and Watson, K. M., *Collision Theory,* New York: John Wiley and Son, 1964, chapter 10.

Index